CONSTRUCTION PROJECT PLANNING AND SCHEDULING

CONSTRUCTION PROJECT PLANNING AND SCHEDULING

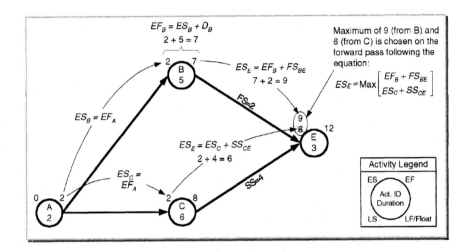

Charles Patrick, P.E., C.S.I.T.

Department of Industrial Education and Technology
Morehead State University
Morehead, Kentucky

PEARSON

Prentice
Hall

Upper Saddle River, New Jersey
Columbus, Ohio

Editor in Chief: Stephen Helba
Editor: Ed Francis
Production Editor: Christine Buckendahl
Production Coordination: Carlisle Publishers Services
Design Coordinator: Diane Ernsberger
Cover Designer: Bryan Huber
Production Manager: Matt Ottenweller
Marketing Manager: Mark Marsden

This book was set in ITC Mendoza Roman Book by Carlisle Communications, Ltd. It was printed and bound by R. R. Donnelley & Sons Company. The cover was printed by Coral Graphic Services, Inc.

Pearson Education Ltd.
Pearson Education Singapore Pte. Ltd.
Pearson Education Canada, Ltd.
Pearson Education—Japan

Pearson Education Australia Pty. Limited
Pearson Education North Asia Ltd.
Pearson Educación de Mexico, S.A. de C.V.
Pearson Education Malaysia Pte. Ltd.

10 9 8 7 6 5 4 3 2 1
ISBN 0- 13-092480-6

PREFACE

This book is appropriate for freshman- and sophomore-level project planning and scheduling courses in construction management, construction technology, civil engineering, civil engineering technology, and related engineering/technology programs. The book is particularly useful as a text for a first or second course in construction project management. The text thoroughly covers the topic of the need for and use of project planning and scheduling in the construction industry. The primary focus of the text is the development of work breakdown structures; precedence grids; precedence network node diagrams; analytical methods for network solutions; resource scheduling, leveling, and allocation; and project scheduling simulation, including the application of PERT.

Chapter 1 introduces the concept of project management from a broad business and industry perspective. This chapter provides a brief history of project management and traces the influence of the development and use of accelerating computer hardware and project management software. Categories of construction projects and the unique nature of construction work are discussed. The application of project management in various types of construction projects is described as well as the need for organized planning and scheduling of construction projects.

Chapter 2 describes the life cycle of a construction project, including the owner's need, conceptual planning, location considerations, design, estimation/ bidding, procurement/construction, and owner occupancy. The work breakdown structure is defined, and the process for developing a WBS is provided as well as the general criteria for an appropriate level of detail. The chapter identifies the key players in the construction process, describes their roles, and shows and characterizes the three common organizational structures for construction projects. Typical construction management techniques and participants during the design and build phases of construction projects are presented.

Chapter 3 details the development of the network plan, including reviewing plans and setting objectives, analyzing the design for alternative construction methods, and developing the network model. The majority of the chapter explains the steps in developing the network model, including defining, sequencing, and diagramming project activities, assigning durations, and calculating schedules. Descriptions of four schedule-diagramming techniques—bar charts, activity on arrow, activity on node, and time-scaled networks—are provided along with advantages and disadvantages of each. The critical path method is explained as the basic methodology used widely to perform project scheduling for projects with particular characteristics.

Chapter 4 describes the construction project in network format—a graphical diagram of the order of events or sequence in a construction project. The network model graphically represents the project typically in activity-on-node (AON) format. The AON network format is used to perform analytical computations by the forward and backward pass method by the critical path method (CPM). These computations determine the project's critical path, the project duration, and the start and finish times of all project activities.

Chapter 5 provides an extension of the critical path method from Chapter 4, called precedence networking (PN), and allows the use of complex relationships between activities including the start-to-start and finish-to-finish relationships with associated lag time. Computations for the PN method are explained and detailed through examples. This chapter includes an overview of Primavera's SureTrak project management software. The chapter explains the use of the software by using an example project.

Chapter 6 presents general resource management in construction projects, including a discussion about construction resources and resource management in the construction industry. The chapter explains the general trends of resource-constrained schedules as well as the analytical methods used for resource management—scheduling, leveling, and allocation. The sum of squares analytical method of resource leveling and series/parallel methods of resource allocation are described.

Chapter 7 explores methods to reduce project duration that minimize project costs. Reasons for the need to reduce project duration are discussed, and the time-cost trade-off process is explained. The chapter discusses methods to reduce project duration without increasing total project costs and explains how time reduction can be achieved by expediting or buying time along the critical path. Included is an explanation of time-cost relationships for project activity duration reduction as well as cost slope computations and comparisons.

Chapter 8 focuses on uncertainty in activity durations and examines a method, the program evaluation and review technique (PERT), to manage this uncertainty. A limitation of the traditional network models (CPM and PN) is detailed—that is, activity duration estimates are fixed and known. The PERT procedure is a probability modeling method for scheduling projects that have highly variable and uncertain activity durations. The chapter outlines a conceptual framework of the PERT process, explains the procedure formulation, and gives several examples.

ACKNOWLEDGMENTS

I thank Mr. Tom Hale, Editor for Associated Construction Publications, *Construction Digest* (Indianapolis), Mr. Steve McGlothen, Project Manager for Manus, Inc. (Lexington), Dr. Wafeek Samuel Wahby, Coordinator of Industrial Technology at Eastern Illinois University, and Mr. Tim Holbrook, Chief Photographer for Morehead State University (Morehead), for their photograph contributions to this text. In addition, I thank Mr. Rob Crouch for the townhouse CAD drawings he generated for the appendix. Thanks also to Mr. Ed Francis, Executive Editor, and Ms. Jennifer Day, Editorial Assistant, for their input and guidance during the writing and editing process.

Thanks, also, to the reviewers of this text for their helpful comments and suggestions: Terry L. Anderson, University of Southern Mississippi; David Bilbo, Texas A&M University; S. Narayan Bodapati, Southern Illinois University, Edwardsville; William Campbell (retired), Montgomery College; Charles W. Graham, Texas A&M University; Ahmed Hadavi, Northwestern University; John Jarchow, Pima Community College; David Leo Lickteig, Georgia Southern University; Liang Liu, University of Urbana, Champaign; and James Stein, Eastern Michigan University.

I am grateful to Morehead State University for granting me a sabbatical leave during the 2002 spring semester to complete the book. Thanks also goes to Mr. Dennis Karwatka, an experienced textbook author and Professor of Industrial Education and Technology at Morehead State University, for guidance and advice during the development and writing of the book. And I thank my students who, for over 20 years, have provided me the motivation to take on challenging ventures.

Most important, I am sincerely thankful to my family—my sons Kyle and Evan, my daughter Anne, and especially my wife Martha. Their patience, support, interest, and advice during this endeavor made the book possible. Finally, I thank my dad Bill Patrick, who left us in April 2000, and my mom Alma Patrick for encouraging me to do my best all my life.

Charles Patrick, P.E., C.S.I.T.

BRIEF CONTENTS

CONTENTS

CHAPTER 1

CHAPTER 2

CHAPTER 3

THE NETWORK PLAN   51

Chapter 6

RESOURCE SCHEDULING, LEVELING, AND ALLOCATION   137

Chapter 7

PROJECT TIME REDUCTION / TIME-COST TRADE-OFFS   161

Chapter 8

PROJECT SCHEDULING WITH UNCERTAIN DURATIONS   177

APPENDIX

PROJECT MANAGEMENT IN THE CONSTRUCTION INDUSTRY

OBJECTIVES

This chapter provides knowledge in the areas of:

- ❑ Project management in business and industry
- ❑ Construction project management
- ❑ Construction industry-projects in construction
- ❑ Unique nature of construction "plant"
- ❑ Need for construction project planning and scheduling
- ❑ Effect of planning on project cost

1.0   OVERVIEW

This chapter introduces the concept of project management from a broad business and industry perspective. The chapter provides a brief history of project management and the development and use of accelerating computer hardware and project management software. In addition, the chapter discusses categories of construction projects and the unique nature of construction work. The application of project management in various types of construction projects is described. One of the most critical sections of this chapter focuses on the need for organized planning and scheduling of construction projects. Finally, the chapter explains the effect of planning on project cost.

1.1   PROJECT MANAGEMENT IN BUSINESS AND INDUSTRY

1.1.1   INTRODUCTION

Project management is an objective-oriented process used throughout business and industry. The **management of projects**, whether conducted formally or informally, incorporates methods to effectively coordinate and utilize project resources for the accomplishment of one or more objectives. Project management is typically carried out under different types of constraints, including time, labor, material, equipment, cost, and performance of the end product. The planning, scheduling, evaluation, and control of these constraints present distinctive management problems varying from mere nuisances to true impediments for getting the project completed.

In the construction industry, construction projects require knowledge of project and business management as well as an in-depth understanding of the construction process. A distinction is made between construction project management and general business management in that project management is used to accomplish specific construction project objectives. Once the project is completed, the construction management personnel focus on an entirely separate large project or on multiple smaller projects. In most instances, this "new" project(s) will have little, if any, relationship to the previous project other than the basic nature of construction work. An example of this is the erection of steel-frame warehouse buildings such as shown in Figure 1.1. Although the previous project may have also been a steel-frame warehouse, the uniqueness of building designs coupled with the unique nature of the construction process at individual sites makes each project relatively "new." The unique nature of the construction process is discussed later in this chapter.

1.1.2   HISTORY OF PROJECT MANAGEMENT

There is evidence that several cultures of humans have practiced project management during at least the last 5,000 years. For example, the construction of the great pyramids (see Figure 1.2) required thousands of workers performing manual labor for several decades. The work on a pyramid was likely overseen not by hundreds, but only a few individuals. These individuals would have been people in Egyptian society educated in mathematics, communication, and literature. They would have been in charge of the entire building process. They had to have had knowledge of project management and the skills to be able apply those principles (Lehner, 224).

FIGURE 1.1 Warehouse steel structure
SOURCE: *Manus, Inc. (Lexington, KY)*

FIGURE 1.2 Giza pyramids
SOURCE: *Dr. Wafeek Samuel Wahby*

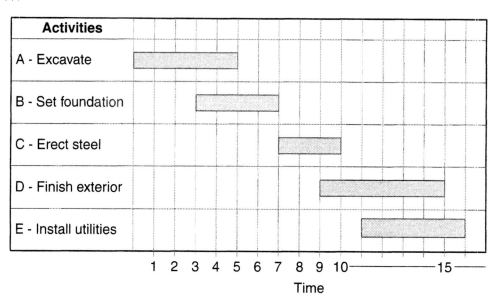

FIGURE 1.3   Classic Gantt, or Bar, Chart Schedule

The more recent history of project management began with increased industrialization during the late nineteenth century. In the United States, the undertaking of the transcontinental railroad was a project that required organizing thousands of workers for unprecedented levels of manufacturing and fabrication. At about the same time, Frederick Taylor applied scientific reasoning to the work process through time and motion studies primarily in the manufacturing industries. Taylor's time and motion studies provided a basis to analyze and improve the work process at the singular worker level. An associate of Taylor, Henry Gantt focused his studies on U.S. Navy ship construction at the early part of the twentieth century. His Gantt, or bar, chart method provided a visual representation of the duration and sequence of project activities. This classic scheduling chart (see Figure 1.3) is still widely used today due to its ability to effectively communicate project progress in a graphical, easy-to-understand format. Taylor's and Gantt's methods continued to be used after World War II. In fact, Gantt's bar chart remains a mainstay of contemporary project scheduling techniques.

1.1.3   CONTEMPORARY PROJECT MANAGEMENT

In the mid-1950s mainly to take advantage of mainframe computers, techniques were developed to manage and control resources (i.e., time, people, money, materials) for complex projects. The techniques included arrow-and-node network diagramming, the critical path method (CPM), and the program evaluation and review technique (PERT).

In the early 1960s, the progression of business and industry resulted in a rising complexity of business and industrial systems. It wasn't just a matter of undertaking larger and larger projects; there existed a need to communicate and integrate work across various departments in the business and industrial environment. At this time, systems analyses began to be applied to business and industry.

Systems analysis is an approach that concentrates study on the constituent parts of complex systems. Unlike the Taylor and Gantt reductionism approaches (i.e., the bottom-up explanation of the whole from the component properties), systems analysis is based on the idea that complex events often result from interactions within the system. A natural result of systems analysis was the project management ideal—organizing work in projects that connect many disciplines and departments existing in business and industry, enhancing communication and efficiency.

One of the most influential developments in the field of project management began in the early 1980s. The development was the growth and rapid advancement of personal computers and project management software. During this time, the level of sophistication in both computer hardware and project management software increased, and so did their ease of use due to enhancements in graphical user interfaces. These advancements in computer hardware and project management software put advanced tools in the hands of project managers, not just computer programmers.

While computer development accelerated during the 1980s, other industries were facing very difficult economic circumstances. The United States and much of the rest of the world saw high interest rates and recession. There was "global pressure for drastically improved quality and shortening of the product development cycle" (Kerzner 1994, 6). Recessions force companies to emphasize efficiency and productivity. The emphasis tends to naturally promote the concepts of careful planning, scheduling, and controlling of projects and thus the project management axiom throughout all business and industry.

The accelerated growth of computers and the recession of the 1980s increased the need for qualified project managers. These project managers were, and are today, expected to possess the technical and people skills to plan, schedule, evaluate, and control complex projects in business and industry using advanced tools (computer hardware and software). To illustrate this, a recent survey of over forty project management professionals in the construction industry indicated that nearly 100 percent used computer-based project management during the previous 12-month period (Liberatore, Pollack-Johnson, and Smith 2001). While this survey was limited to only forty project management professionals and may not accurately represent that entire population, the survey conclusions are reasonable and tend to indicate an increasing need for computer-based knowledge and skills enhancement in the project management field.

1.2   CONSTRUCTION PROJECT MANAGEMENT

According to the Project Management Institute, the discipline of project management can be defined as follows:

> Project management is the art of directing and coordinating human and material resources throughout the life of a project by using modern management techniques to achieve predetermined objectives of scope, cost, time, quality and participation satisfaction. (Wideman 1986, (16)).

This definition of project management is widely applied in business and industry. For our purposes, a more narrow definition is needed to focus attention on the application of project management to the construction industry.

Many authors from various academic fields have contributed to the research and literature on project management. This is also true in the field of **construction project management.** Feigenbaum (1998) defines project management as "the comprehensive process of planning, directing, and controlling the construction project from its inception through its completion" (p. 2). Marchman (2000) states that "good project management means 'project controls' rather than a project 'out of control'" (p. 6). He gives the purpose of project management as the achievement of "a project's goals and objectives through the planned expenditure of resources that meet the project's quality, cost, time, and safety requirements" (p. 7). Moder (Moder, Phillips, and Davis 1983) explains that "project management involves the coordination of group activity wherein the manager plans, organizes, staffs, directs, and controls to achieve an objective with constraints on time, cost, and performance of the end product" (p. 3). Spinner (1997) defines project management as "managing and directing time, material, personnel/labor, and costs to complete a project in an orderly, economical manner and to meet the established objectives of time, costs, and technical and/or service results" (p. 4). Spinner goes on to include planning, scheduling, and controlling as the three stages of successful project management.

These definitions provide a general overview of project management. To more accurately define project management as applied in the construction industry, a definition of each term—*project* and *management*—is useful. A **project** is a group of discernible tasks or activities that are conducted in a coordinated effort to accomplish one or more unique objectives. Projects require varying levels of cost, time, and other resources (i.e., labor, equipment, material, supplies, tools). A project is usually distinguished from other repetitive or continuous processes such as occurs in a manufacturing line. Projects are found throughout business and industry in areas such as manufacturing facilities, new product or product line development for business, infrastructure development and improvement, and residential and commercial building (as shown in Figure 1.4).

FIGURE 1.4 Commercial building project
SOURCE: *Construction Digest*

A universal definition of business **management** is to effectively plan, organize, staff, direct, and control a process for the achievement of goals and objectives. More specifically, construction management includes the steps of **planning, scheduling, and controlling (or monitoring)** the various phases or segments of a construction project. *Planning* a construction project is an essential activity in which techniques are selected, project tasks are defined, and resource requirements and constraints are identified. Construction planning prepares for the efficient application of project resources. *Scheduling* is arranging project tasks and resources in a sequence to minimize the use of resources, specifically time and money, and maximize quality and customer/owner satisfaction. *Controlling (monitoring)* is the process of comparing scheduled and actual events and, after analyses, correcting the difference by coordinating the action of all parts of the project team.

Given these definitions, **construction project management**, or project management applied in the construction industry, can be defined as follows:

> *Construction Project Management:* The planning, scheduling, evaluation, and controlling of construction tasks or activities to accomplish specific objectives by effectively allocating and utilizing appropriate labor, material, and time resources in a manner that minimizes costs and maximizes customer/owner satisfaction.

1.3 CONSTRUCTION INDUSTRY—PROJECTS IN CONSTRUCTION

As defined by total annual expenditures, the construction industry is one of the largest industries in the world. The industry is an aggregation of diverse trades and professions that have a strong influence on the global economy. The diversity is very evident, particularly in technological societies, in the number and types of residential, commercial, industrial, and heavy construction throughout the world.

The primary activities of the industry are the construction of new facilities and the renovation of existing structures. Construction projects are undertaken because an owner (private or public) needs or desires a facility (e.g., house, building, road, airport). All constructed facilities are built following the intent of function and form. The facility must meet certain functional demands to satisfy discrete objectives within specifications defined by the owner as well as legal and regulatory constraints. The facility form (design, appearance, pattern, color, etc.) must satisfy the owner's need for uniqueness and individuality as well as applicable zoning regulations.

For the purpose of describing diverse projects in the construction industry, all construction work can be categorized under two main headings: **heavy (or horizontal) construction** and **building (or vertical) construction**, comprising residential and commercial (or nonresidential) construction. Heavy construction projects make up the infrastructure system of a city, county, state, and nation. Examples of heavy construction are roads, bridges (as shown in Figure 1.5), dams, airports, and tunnels. Most of these projects are funded and monitored by appropriate government agencies, using public funding, and are built by heavy construction contractors.

In building construction, *residential* describes private housing and *nonresidential* describes commercial buildings, although these definitions must be applied broadly. Residential house construction tends to be privately funded by

FIGURE 1.5   Bridge construction project
SOURCE: *Construction Digest*

individual owners, and general building contractors are the primary builders. Nonresidential construction includes office buildings, large housing complexes, malls, churches, theaters, and schools. Many of these projects are privately funded although some projects (e.g., public housing, schools, and courthouses) are publicly funded. Typically, general contractors with experience in nonresidential construction build these projects. An example of a commercial construction project, a stadium, is shown in Figure 1.6.

Large, complex projects sometimes fall into more than one of the above categories. Industrial plants, steel mills, and manufacturing factories commonly require a great deal of earthwork (heavy construction) prior to erection

FIGURE 1.6   Construction of a stadium
SOURCE: *Construction Digest*

of the physical plant (building construction). Large pipeline projects (e.g., gas, oil) require linear earthwork similar to that needed for road building and pipe placement (heavy) accompanied by transfer stations and processing plants (building). Because of the variety of equipment types and skilled labor, as well as management expertise, required, few contractors have the design and/or build capabilities to complete an entire project of this magnitude. Large, complex construction projects are commonly undertaken by design-and-construct or construction management firms that organize, direct, and control the necessary designers, contractors, and resources (time, materials, labor, and equipment) to complete a large project.

1.4   UNIQUE NATURE OF CONSTRUCTION PLANT

Individuality is a desirable human characteristic, especially with a growing population and expanding global society. The global society is, in part, more prevalent due to significant enhancements in modern technology. Yet the extensive variety in technology, particularly material goods, is one of the principal ways people express their individuality. In the United States alone, millions of new automobiles are bought each year. When purchasing a new vehicle, people tend to look for something unique—something different. A recent advertising campaign for Dodge/Chrysler is based on their "different" vehicles. Being different or unique is the key selling point for many products.

This desire for individuality is no more evident than in the vast diversity of structures that humans build. This can be seen in our homes, churches, schools, office buildings, airports, and bridges. Even in large developments where houses have very similar interior layouts, people put much effort into making their house look different, particularly the exterior. Why? It's simple human nature to want a house different from others in the neighborhood—it's an expression of our individuality.

Successful designers or architects base their success, at least partly, on their ability to design structures unlike those built before, whether a traditional ranch house, long suspension bridge, or towering skyscraper. Subsequently, the construction industry must interpret these unique designs and build the structures to meet the owners' individual needs. These structures are the products of the builders, and these products must be fabricated using a specific process. The uniqueness of design and resulting construction is clearly evident in projects like the William H. Harsha cable-stayed bridge crossing the Ohio River between Maysville, Kentucky, and Aberdeen, Ohio. A towering pier of that bridge is shown in Figure 1.7.

The construction industry is similar to the manufacturing industry in that a product is fabricated using a specific process. Both industries set up at a particular site, select and procure the appropriate resources (materials, equipment, and labor), and fabricate the product. Both industries are interested in producing a high-quality product in the most efficient manner, normally in the shortest time possible.

There are, however, two primary differences between the construction and manufacturing industries. In manufacturing, the plant location is normally **static**, or in one location that does not change, particularly during the fabrication of a specific product. A manufacturing company commonly stays in a production facility or plant for an indefinite time. A variety of products may be manufactured, but the plant infrastructure (environmental conditions, equipment, tools, bathrooms, parking, administrative and training offices) stays somewhat unchanged. In the construction industry, the "plant" location (building site) is normally **dynamic** and changes with each fabrication of a particular product. No two structures, regardless of similarities, are built at exactly the same site or plant

FIGURE 1.7 Cable-stayed bridge pier, Maysville, Kentucky
SOURCE: *Construction Digest*

(see Figure 1.8). In addition, each time a product is built, the plant infrastructure is not likely to be the same as the last build. For example, if two similar structures are built adjacent to one another, one from June to October and the other from November to February, the building process is likely to be altered to accommodate, at a minimum, the weather differences. In fact, environmental factors—weather, topography, and soil conditions—can and will change significantly from one job to the next.

The second key difference between the manufacturing and construction industries is in **process standardization**. Two key objectives of manufacturing are to optimize product quality and productivity of the manufacturing process. These are desirable objectives in the construction industry as well, but they are

FIGURE 1.8   Concrete testing on a bridge deck
SOURCE: *Construction Digest*

attained in manufacturing by exact repetition of individual steps in the fabrication process for a particular product. While a wide variety of products may be fabricated in a manufacturing facility and sometimes in low numbers, high volumes of single products are not uncommon and are used here for comparison purposes.

For a particular product, the manufacturing process is carefully analyzed in order to develop a specific flow of work. This analysis results in process flow charts, which detail workflow and tasks. Generally, line or floor workers are trained to make a product at high volumes and quality by repeating the production process exactly the same way for each iteration of the process. This system is designed to allow line workers to get to a point where they know their particular task in the process so well that it becomes standardized for them and high volume and quality are possible. Process standardization is a crucial characteristic of successful manufacturing plants.

The construction process is more difficult to standardize. The general phases of construction projects are somewhat standardized (e.g., home building, road building) and follow a general sequence of ordered activities. A good example is the structure of a commercial building, where the foundation must be complete prior to erecting the frame and roof (see Figure 1.9).

Workers gain a standard, general knowledge and skills of the building process from experience in similar "builds." However, there is no set of standard sequences for these processes because every construction project and thus resulting product is unique. Each job (e.g., house, building, road, airport) differs in its function and form, particularly form. To some degree, no two jobs are ever quite

FIGURE 1.9   Commercial building foundation, frame and roof
SOURCE: *Manus, Inc. (Lexington, KY)*

alike. Therefore, the process of fabricating a product in the construction industry can be classified as a **nonstandardized process.**

> The construction industry differs from other heavy industries and is thus unique due to the *dynamic plant* and *nonstandardized nature of the process.*

1.5   NEED FOR CONSTRUCTION PROJECT PLANNING AND SCHEDULING

Throughout human history, individuals and societies have grown ever more sophisticated and complex. As humans evolved, so did the structures where they lived, worked, and played as well as the structures that supported their societal infrastructure. Construction processes also evolved as humans gained experience and knowledge about building. Over time, many diverse and grand structures were constructed including the Egyptian pyramids, Roman coliseums and aqueducts, and European cathedrals. The evolution of structures and building processes accelerated during the twentieth century and continues today. Key technological and societal developments have influenced this accelerated evolution, including mechanical power improvements, advancements in metallurgy and composite materials, improved analyses of building processes, rising human knowledge and

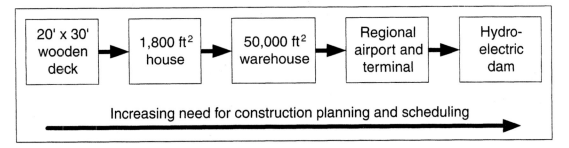

FIGURE 1.10   Effect of project size on need for planning and scheduling

skill levels, advancing educational systems, and an ever increasing demand for goods and services.

During the twentieth century, many remarkable construction projects were completed, such as the Golden Gate Bridge, Chunnel (or Euro-tunnel, the transportation tunnel beneath the English Channel), Alaskan pipeline, Japan's Akashi Kaikyo suspension bridge, Sears Tower skyscraper, and Hoover Dam. Both the design and construction processes used to build these impressive structures required innovation and significant resources—time, materials, labor, equipment, and wealth. For each of these building projects and all others over time, regardless of size, construction planning and scheduling was part of the process.

One can only imagine the challenge of planning and scheduling the Great Pyramids of Giza, as shown in Figure 1.2, before and during a building process that lasted decades. However, even the smallest structure, like a simple residential porch or deck, must have some level of planning and scheduling. It may be a simple matter of one person thinking carefully about the building process and the required resources (i.e., time, materials, and tools) for the structure.

As the complexity of projects increases, as illustrated in Figure 1.10, the requirement of an **organized planning and scheduling process** is enhanced. An organized planning and scheduling process would likely include a sequential list of project tasks at a level of detail that would effectively communicate project times, responsibilities, and costs to the appropriate participants. Further details of specific contents for and development of the planning and scheduling process are explained in later chapters.

The need for organized planning and scheduling of a construction project is influenced by a variety of factors. Several of the more significant factors include:

❑ Project size and number of project activities
❑ Project type
❑ Background of the project/construction manager
❑ Contract requirements

❏ Owner/customer appeal
❏ Resource analysis
❏ Risk assessment

1.5.1   PROJECT SIZE AND NUMBER OF PROJECT ACTIVITIES

When erecting a structure, a sequential or ordered construction process must be carried out. Certain activities must occur either before (predecessor) or after (successor) others. For example, the foundation for a warehouse must be finished before the frame and roof are started; concrete for a retaining wall cannot be poured until the forms are complete. As project size and complexity increase, so does the need for organized planning and scheduling (Figure 1.10). The project size directly affects the number of physical steps and the amount and kinds of material and equipment that are typically needed to complete the project as well as the number and variety of skilled workers required. These additional resources must be managed more carefully to accomplish the project on time and within budget. Thus, more attention must be paid to planning and scheduling these resources.

To plan and schedule a construction project, activities must be defined sufficiently so that adequate communication is provided to all those who will use the information. Participants in planning and scheduling will likely include the owner(s), construction supervisors, workers, inspectors, and others. Not everyone will need the same information; some will need more detailed planning and scheduling. This **level of detail** determines the number of activities contained within the project plan and schedule. As the number of project activities increases and thus the complexity of their sequential ordering, the need for organized planning and scheduling increases. This need further increases when a large number of project activities are considered relative to the uniqueness of each construction project in terms of the dynamic plant and nonstandardized nature of the work. The level of detail for defining project activities is further discussed in Chapters 2 and 3.

A relationship exists between project size and number of activities, as represented in Figure 1.11. Small projects require a relatively small number of activities. As project size increases, so does the number of required activities, but slowly at first because the project does not generally get more complex and the planner simply adds more activities. The lower one-third portion of the curve in Figure 1.11 represents this. As project size continues to increase, project complexity also increases (i.e., sequential ordering, activity relationships) and thus the number of activities representing the project increases more rapidly—the middle portion of the curve. This continues until a large number of activities are required to represent the entire project. At some point, the number of activities begins to become unmanageable for the project planner, slowing the rate of growth—the upper portion of the curve. As the project reaches very large proportions, the project planner will most likely break it into more manageable phases of construction, with distinguishable points of completion, or **milestones**, delineating the end of one phase and the beginning of another.

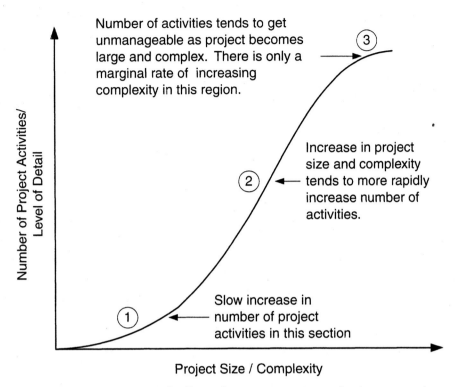

Number of activities tends to get unmanageable as project becomes large and complex. There is only a marginal rate of increasing complexity in this region.

Increase in project size and complexity tends to more rapidly increase number of activities.

Slow increase in number of project activities in this section

Number of Project Activities/ Level of Detail

Project Size / Complexity

FIGURE 1.11 Conceptual effect of project size/complexity on number of defined project activities

1.5.2 PROJECT TYPE AND MANAGER BACKGROUND

Certain types of construction projects tend to have more complex activity relationships and dependencies than others as the complexity of the project increases. For example, commercial construction projects will be more complicated in terms of erection techniques and resource management than will residential projects. It may be difficult and possibly overwhelming for one person to keep all the project details in his or her head and recall those details from memory. The project or construction manager's individual work experience with the particular type of project, educational background, and planning and scheduling expertise will also influence the need for organized planning and scheduling. A project or construction manager highly experienced in residential construction may need very little organized planning and scheduling for a 1,800-ft^2 home. After doing several similar projects in the past, the process gets somewhat standardized for the manager. On the other hand, few people would attempt to plan and schedule a large project (e.g., multistory office building, airport, or dam) without an organized approach.

The planning process may be one where time is considered a critical resource, such as in the case of industrial facility construction. If the owner, for example a manufacturing company, is trying to capture a certain market for a product from a new industrial facility, demand for the product may decrease over time. Thus it is crucial for the facility to be built quickly. On the other hand, an owner wanting a new residential house built is less sensitive to variations in a scheduled deadline. While the owner expects the contractor to stick to a certain predetermined schedule and a delay of an extra month in the build will affect the contractor's reputation and possibly delay the start of another job, the final product will not lose its demand. In other words, the owner will still likely want the house.

1.5.3   CONTRACT REQUIREMENTS

The owner funding a construction project can be either private or public. While written contracts are not always used in private projects (although recommended), public projects always include a detailed contract between the owner(s) and prime contractor. Many contracts specifically require construction contractors to provide a detailed time schedule for the project prior to the actual beginning of the work. This typically must include a sequentially ordered set of project activities shown in graphical format. In some cases, the time schedule must be submitted with the bid or cost estimate.

1.5.4   OWNER/CUSTOMER APPEAL

From a business perspective, construction contractors are in business to provide a service to the public and make money. Maintaining costs and producing a quality product accomplish this. Good planning and scheduling help the contractor better manage project resources (time, labor, materials, equipment, costs). Well-managed projects allow the contractor to complete a project within budget and to a certain level of performance and quality. The contractor who provides a quality product within the budgeted cost and time not only accomplishes the project goals but also appeals to the owner/customer and thus increases the likelihood of future profits through repeat business and referrals.

1.5.5   RESOURCE ANALYSIS

The analysis of resources, particularly time, materials, labor, and equipment, is key to good project management. Project planning and scheduling allows the contractor to preview and lay out the project—activities, sequential relationships and dependencies, and activity and project duration. The contractor can determine the sequence of activities that are most critical for the successful completion of the project and allocate the appropriate resources for this sequence. Of great importance is the ability to determine and make adjustments for resource conflicts, especially for those activities that are most critical. The coordination of resources over the length of the project to meet deadlines and quality requirements, as defined by contract documents, helps the contractor complete the project within budget and on time.

1.5.6 RISK ASSESSMENT

All activities within a construction project carry some level of risk or uncertainty in terms of cost and time. Project activities may be delayed and take longer than expected to complete. Procurement of materials, supplies, and/or equipment may not adhere to schedule. The likelihood of these unexpected events, or risk, is influenced by the changing environmental conditions in the dynamic construction plant, the unique form and function of each construction project, and the non-standardized nature of project work. The contractor wishes to finish the project activities and thus the project as quickly as possible while maintaining acceptable quality. Shorter project times tend to reduce costs for the contractor and increase profits. Therefore, it is best for the contractor to perform detailed planning and scheduling of the project to minimize unexpected events. The construction/ project manager must concentrate on those project activities that have the most risk or uncertainty, just as an estimator would concentrate most of his or her effort on the most risky items in a bid.

1.6 EFFECT OF PLANNING ON PROJECT COST

A construction project begins because an owner, whether an individual, corporation, government organization, or other entity, needs or desires to have a facility or structure built. The owner typically contacts a designer or design company, which turns the owner's ideas into working plans (i.e., drawings and specifications). A contractor is chosen, through competitive bidding or negotiation, to build the facility. In this process the designer—and, in some cases, the contractor (after awarded a contract)—assists the owner in planning the project to set the overall project objectives of cost and time. The contractor provides detailed cost estimates and time schedules to the owner. A thorough explanation of this owner-designer-contractor process is provided in Chapter 2.

The designer's and contractor's estimates provide the owner the probable cost of the project. This is the capital project cost and does not include the operation and maintenance cost of the facility after the owner has taken occupancy. The capital cost of a project involves several elements:

❑ Land acquisition
❑ Design of the facility including:

- Development of project objectives, initial cost estimates, and initial time schedules

- • Generation of working drawings, specifications, and contract documents
- ❏ Construction equipment and labor
- ❏ Materials for the facility
- ❏ Construction planning—contractor's time to plan and schedule necessary resources
- ❏ Management (control) during facility construction
- ❏ Overhead, insurance, taxes, etc.

Of these various elements, construction costs tend to be the largest portion of the total project cost regardless of project type or uniqueness. In addition to the obvious construction costs for materials, equipment, and labor, the costs of planning and scheduling resources as well as field supervision are important components. The contractor must carefully plan in order to finish the project on time and within budget. Thorough planning prior to the start of construction is critical to minimize costs because *any decision related to design and/or resource management made at the beginning of a project will cost much less than those made as the project progresses, especially in late stages of the project.* This concept is illustrated in Figure 1.12.

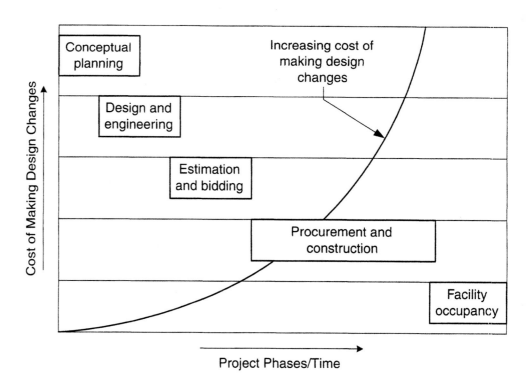

FIGURE 1.12   Effect of planning on project cost

1.7 SUMMARY

Project management incorporates methods to effectively coordinate and utilize project resources for the accomplishment of one or more objectives. Project management is typically carried out under some level of constraint for the project factors of time, labor, material, equipment, cost, and performance of the end product. An organized, scientific approach for project management began with Frederick Taylor and Henry Gantt in the early twentieth century. Taylor and Gantt set the stage for the development of various analysis techniques in project management including arrow-and-node network diagramming, the critical path method (CPM), and the program evaluation and review technique (PERT). The accelerating development and use of computer hardware and software since the late 1980s put sophisticated project management tools in the hands of construction managers.

Construction projects can be categorized as residential (homes), commercial (e.g., office buildings), heavy (e.g., roads and bridges), industrial, or some combination of these. The construction industry differs from other heavy industries due to the dynamic plant and nonstandardized nature of the process. Project management is applied in the construction industry to plan, schedule, and control construction tasks or activities to accomplish specific objectives by effectively utilizing appropriate resources in a manner that minimizes costs and maximizes customer/owner satisfaction. The need for project management in construction—and more specifically an organized planning and scheduling process—increases as projects get larger and more complicated. An organized planning and scheduling process would likely include a sequential list of project tasks at a level of detail that would effectively communicate project times, responsibilities, and costs to the appropriate participants.

Several factors influence the need for the organized planning and scheduling of a construction project, including the project size, number of project activities, project type and the background of the project/construction manager, contract requirements, owner/customer appeal, resource analysis, and risk assessment. Planning and management of resources have a significant impact on the cost of a construction project. Preliminary planning prior to the actual start of a project is critical to minimize project cost.

CHAPTER 1 QUESTIONS/PROBLEMS

1. Think about and plan for a possible career as a project manager in the construction industry. Develop an objective, sequential list of steps you will need to take to move into and advance in this career path. Include at least seven steps.

2. Develop a list of activities or tasks to build a 12-ft by 12-ft single-level deck set against a residential house. Assume that one skilled carpenter and one

helper will do the work and all materials and tools are readily available. Arrange the project activities in a sequential order.

3. How is project management in general business similar to project management in the construction industry? How are they different?

4. What is the meaning of the term *dynamic construction plant*?

5. Describe what the *nonstandardized process* in the building process means, as compared with what it means in the manufacturing industry.

6. Describe three crucial factors that increase the need for organized project planning and scheduling in the construction industry.

7. Describe, in narrative format, a simple example in the construction industry illustrating the concept in Figure 1.12 in this chapter.

REFERENCES

See References on page 199.

CONSTRUCTION PROJECT MANAGEMENT ORGANIZATION

This chapter provides knowledge in the areas of:

- ❏ Construction project stages
- ❏ Work breakdown structure
- ❏ Example Project- Three-Unit Townhouse
- ❏ Key players in the construction process
- ❏ Construction project organizational structures
- ❏ Design-phase construction management
- ❏ Management of field construction

2.0   OVERVIEW

This chapter begins with a description of the life cycle of a construction project, including the owner's need, conceptual planning, location considerations, design, estimation/bidding, procurement/construction, and owner occupancy. The work breakdown structure, or WBS, is defined, and examples are given. The chapter presents the process for developing a WBS as well as the general criteria for an appropriate level of detail. Next, the chapter introduces the three-unit townhouse example project that is referenced throughout the text. The owner, designer, and contractor are identified as the key players in the construction process and their roles described. The three common organizational structures for construction projects are shown and characterized. Finally, the chapter discusses typical management techniques and participants during the design and build phases of construction projects.

2.1 CONSTRUCTION PROJECT STAGES

The stages of a construction project, as described here, include all steps in the process, from the initial market demand or perceived need of the owner to the contractor turning over the finished facility (Figure 2.1). This comprehensive process is sometimes referred to as the project life cycle. Several key players are involved in the process, including the owner, designer, and contractor. The descriptions and responsibilities of these key players are described later in this chapter.

2.1.1 CONCEPTUAL PLANNING AND LOCATION

Construction projects are undertaken because an owner needs or desires a facility to be built. The owner—the contractor's customer or client—will be private or public, probably an individual, family, corporation, organization, or government enterprise. Most likely, the owner will have some idea about the type of structure needed in terms of function and form to satisfy certain constraints. The degree to which the owner is reasonably sure about the use and appearance of the facility depends on the owner's level of experience with previous structures and designers (architects/engineers) and construction contractors.

The owner, at this point, has a general idea about a site or location for the facility or has actually selected a location. The location directly affects the cost of the facility, the construction methods used to erect the facility, and the amount of site preparation needed to begin the facility. The cost of the site depends on land use and adjacent property values and will likely involve site surveys/plots and land ownership transfers. The construction methods used and the necessary site preparation depend on environmental conditions at the site (see Figure 2.2).

The facility location must meet certain basic requirements. The site must be located such that equipment and materials can be brought to the site without too much difficulty and that workers and subcontractors can access the site also without significant difficulty. In addition, skilled workers and subcontractors must be available in the region or must be moved in at an increased cost. Utilities, including water for drinking, electricity, and heating/cooling, where applicable, must be available. Environmental and physical conditions at the site, including normal weather patterns, the chance of severe weather, topography, and soil types, must be considered. In certain locations, particularly near heavily populated areas, zoning and building codes at the municipal, state, and federal level may affect the

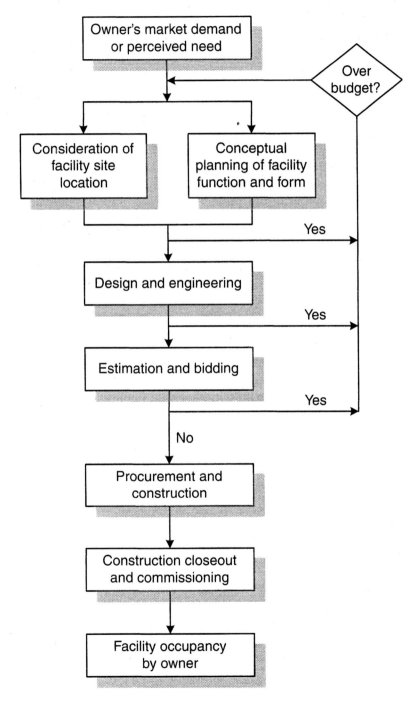

FIGURE 2.1 Construction project stages

Figure 2.2 Installing drains on a construction site
SOURCE: *Construction Digest*

location and the building process. A congested intersection as a result of a construction renovation of the ADUC student center in the center of the Morehead State University campus is shown in Figure 2.3.

2.1.2 Design

Once a reasonably clear idea about the prospective facility and location has been established, the owner can hire one of three types of firms—a design/construct (often referred to as design/build) company, or a project design company, or a construction management agency. A design/construct company has the capability to both design the project with in-house architects/engineers and physically erect the facility as the general contractor. The project design company has the architectural and engineering capability to design the project, and a separately hired contracting company will erect the facility. The construction management agency typically hires design and construction firms and manages all portions of the project. While these three types of firms provide services for a wide diversity of construction projects, it is not unusual for project size to be a determining factor in the choice of firm type. Often, as project size increases, the choice of firm type goes from the design/construct company to the project design company to the construction management company.

During the design, the architectural and engineering aspects of the facility are developed. They include the ability of the structure to meet the needs of the owner, the shape and appearance of the physical structure, and the con-

FIGURE 2.3   Intersection congestion in front of the ADUC student center
SOURCE: *Tim Holbrook, Photographer—Morehead State University*

struction process to be used. It is during this time that cost estimates of the facility are developed as the owner's wishes are communicated to the designers. These cost-of-construction estimates are performed by the architect and/or engineer based on drawings and specifications. The estimates are provided only as conceptual or rough estimates, not actual construction costs. The owner goes through an iterative process of considering facility function/form cost versus the initial budgeted cost the owner originally had in mind.

The architect/engineer will next develop drawings of and specifications for the facility. Drawings are used to communicate ideas to the builder regarding the facility layout, structural elements, electrical/mechanical installations, and some detail on construction techniques (how the facility is to be made). Drawings are also very helpful to the project manager to develop plans and schedules for the building process. Specifications are written statements that inform the builder what is to be built, what materials are to be used, and how the finished facility is expected to perform. Specifications tend to be broken into the categories of legal provisions (documents for bid advertisement and zoning code constraints), general provisions (conditions and responsibilities of all people involved in the project), and technical provisions (the kinds of materials and processes to be used).

2.1.3   ESTIMATION AND BIDDING

For the design/construct company, the procurement of necessary resources and the actual construction (erecting the structure) shift to the construction division of that company. Otherwise, a general contractor is selected through a competitive bidding or negotiated contract process. Owners of privately funded projects have the leeway to select a contractor through either bidding or negotiation. Public construction projects are publicly funded, and thus the general contractor is routinely selected by competitive bidding, sometimes referred to as the low-bid process. Projects are publicly advertised to notify interested bidders; sometimes qualifying contractors are invited to bid. Bidders are informed of the project details (drawings and specifications), develop their estimate, and submit their bid by the specified deadline.

In the competitive bidding process, accurate estimates that minimize risk are critically important for the survival of contractors, particularly those who do most of their work from bids. Competitive bidding requires the contractor to determine the probable costs of the construction project. The contractor estimates these costs by carefully considering cost components, quantity take-offs, and construction procedures for elements such as earthwork, materials, labor, equipment (depreciation), overhead, and profit, as well as subcontractor quotations.

In order to reduce the risks of unscheduled project delays and resource conflicts, project work is broken down into manageable tasks or activities, preliminary planning and scheduling of these tasks are performed, and the estimator uses these data to predict costs. The completed estimate is submitted to the owner (or representative) in the competitive bidding process. Once again, the owner must compare accurately estimated project costs with the conceptual or rough estimates provided during the design phase. If the accurate estimates are above a limit set by the owner, a redesign and second competitive bidding process may be necessary, with the owner realizing the increased costs of the redesign. A simpler alternative may be for the owner to negotiate with the low bidder for reasonably minor changes (e.g., less expensive materials) in order to lower the estimate.

After bids are received from contractors, the owner (and possibly the designer and construction manager) reviews the bids for accuracy and completeness. Of all qualifying bidders, the low bidder is normally awarded the contract. This low bid gives the owner an accurate estimate of the actual project cost. The owner, acting alone or more commonly with the assistance of a design or construction management firm, develops a contract with the contractor chosen to build the facility, generally called the prime contractor. At this point in the process, project drawings and specifications are submitted to the appropriate municipal, state, and/or federal inspector of facilities. This person(s) checks the documents to see if they conform to pertinent building codes and zoning ordinances prior to issuing a building permit.

2.1.4 PROCUREMENT AND CONSTRUCTION

The prime contractor, whether working on a private or public contract, begins the building process with project mobilization (i.e., move in) and procurement. Project mobilization involves setting up the temporary infrastructure [i.e., field office (see Figure 2.4), electrical, and communication] at the building site so that the

Figure 2.4 Field office trailer at construction site
SOURCE: *Manus, Inc. (Lexington, KY)*

contractor can monitor and control the building process. For procurement, the contractor acquires or gathers together the required resources to begin the physical erection of the contracted facility. Resource procurement continues as needed throughout the project until the facility is completed according to contract specifications. It is the responsibility of the prime contractor to select subcontractors to perform specialty functions of the work (e.g., electrical, plumbing, HVAC, landscaping). The type and number of subcontractors selected depends on the prime contractor's capabilities, the available time, and the potential cost savings of hiring subcontractors.

The prime contractor uses the contract drawings and specifications as well as the preliminary project schedule to develop a detailed work plan and activity schedule for the project. The prime contractor and subcontractors perform the work according to this schedule and monitor work progress in the process. Necessary adjustments are made to compensate for any unscheduled delays, material and equipment availability problems, and/or unexpected environmental impediments.

2.1.5 Occupancy

Upon physical completion of the facility, the owner or the owner's representative (i.e., the architect/engineer or construction manager) inspects the facility for completion according to contract documents. In fact, inspections occur

FIGURE 2.5   Road project construction site
SOURCE: *Construction Digest*

frequently during the build to be sure that work conforms to the project plans and specifications. During the final inspection, the owner and/or his or her representative will prepare a **punch list,** or list of items of work requiring immediate corrective or completion action by the contractor. Once this work is completed, a final inspection is conducted. When the facility has been accepted, the owner can occupy the facility for its intended purpose. In some instances where the need is urgent, the building/project may be occupied with "substantial completion" before final acceptance, as in the road project depicted in Figure 2.5.

2.2   WORK BREAKDOWN STRUCTURE

The early stages of a construction project require the setting of project objectives to determine the project cost, facility function and form, and project start and completion dates. These objectives also provide for the efficient management of the project during construction. The development of project objectives begins with the owner's conceptual planning of the facility and the design and engineering portion of the project (Figure 2.1). These initial objectives, while broad at first at least in the owner's mind, must evolve into further detail and be finalized in order to conceptualize and plan the complete project. At this stage, the designer normally guides the owner, who wants the most of the facility in terms of function and form for the optimal cost. The designer provides the owner with conceptual or rough estimates and alternatives for the various components of the facility. At some point, they come to an agreement on the design. The objectives of the project are refined during this iterative process.

Once the objectives have been fully developed, the project must be divided into manageable sets of project tasks or activities. This is necessary to produce project drawings and specifications, to generate a detailed cost estimate for the project, and to schedule and control resources (e.g., materials, labor, equipment, and time) for all project activities. The method typically used to break down the project into manageable components or work activities is called the **work breakdown structure (WBS).** The WBS is a structured approach to enable:

❑ Project planning and scheduling
❑ Performance management and control
❑ Cost analyses

The WBS is widely used in the construction industry because it provides a method to functionally divide a construction project into manageable and ordered parts. As stated in Chapter 1, each construction project is somewhat unique due to use and appearance differences of structures, the dynamic construction plant, and the nonstandardized nature of the construction process. This uniqueness limits the ability of planners to simply copy the construction process of a previous project. Each project must be broken down individually, and the WBS is the most widely used method to accomplish this task. The WBS provides a clear understanding of the work activities to be conducted and of the way those individual activities fit into the overall construction process.

In the explanation of the work breakdown structure, let's look at how the method is defined by a few authors from contemporary sources. The Project Management Institute (2001) defines the work breakdown structure as "a deliverable-oriented grouping of project elements that organizes and defines the total scope of the project. Each descending level represents an increasingly detailed definition of the project work" (p. 75). Feigenbaum (1998) states that the WBS "is a hierarchical diagram that shows how each of the subprojects are tied to a larger subproject—and ultimately how they comprise the entire project" (p. 7). Naylor (1995) explains the WBS as "the way a project can be divided into its basic construction elements and then how to concisely display the whole project" (p. 11). Moder, Phillips, and Davis (1983) state that the WBS is used "to show the hierarchy, or levels, of tasks with a project, and the definition of 'work packages' at the lower or basic levels of work" (p. 138). The following definition of the work breakdown structure incorporates the common thread or theme of the quoted authors.

The *work breakdown structure (WBS)* is a hierarchical system that represents the construction project in increasing levels of detail to define, organize, and display the project work in measurable and manageable components.

The WBS hierarchical structure is a system where project work is classified according to various criteria into successive levels or layers. A practical approach to understand the WBS is to think of it as an upside-down tree; it is sometimes referred to as a *reverse-tree* method (shown in Figure 2.6). Generally, higher levels of the WBS represent decision points or comprehensive statements of project activities, while lower levels define actual work activities. At the top of the WBS,

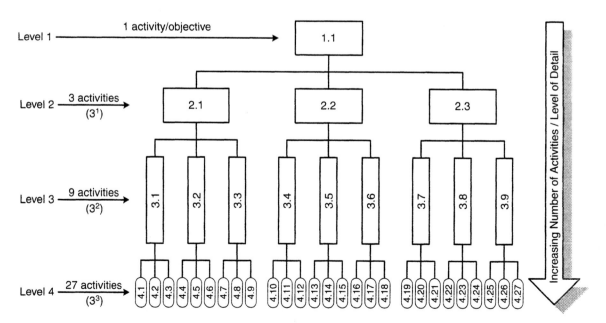

FIGURE 2.6 Conceptual schematic of the work breakdown structure (WBS)

the entire construction project is represented by a single encompassing objective statement (e.g., to build a residential home, to improve an industrial park infrastructure, to erect a warehouse building). The single objective is then broken down into multiple activities typically categorized into specific work groups. Responsible organizational units, physical work areas, subcontractors, or individuals typically categorize these work groups depending upon the size and complexity of the construction project and the management functions of the contractor or construction management firm. The increasingly detailed structure of the WBS aids the management function of the project with the ability to focus attention on and clearly identify responsibility for work activities. This is accomplished at different levels of management through its hierarchical features.

As can be observed in Figure 2.6, the number of activities/objectives in the WBS can grow quite rapidly. In fact, under the assumption that each subordinate level after the first level has only three activities, the second level then has three activities (3^1), the third level has nine activities (3^2), the fourth level has twenty-seven activities (3^3), and so on. While it is unlikely that each activity will have exactly three subordinate-level activities for any project, this example illustrates how the WBS can get large and complex. On the other hand, it's not unusual for a large construction project to be broken into several hundred activities for planning and scheduling purposes. Thus, while a large WBS may be complex and cumbersome, it nevertheless is surely possible.

The development of the WBS is an iterative process that considers the overall purpose of the project, design requirements, performance stipulations, and technical and logistical constraints. This process can be illustrated by a simple example—the building of a relatively small residential structure. The overall objective to "build an 1,800-ft^2 residential structure" is divided into the major building activities of the structure (i.e., foundation, framing/structure, mechanical/electrical systems, and finishes) to create the second level of the WBS (see Figure 2.7). The second level of the WBS is expanded into the third level by dividing each of the second-level activities into specific steps that must be accomplished in order to complete the second-level activities. As can be seen in Figures 2.6 and 2.7, the activities at each level of the WBS follow one simple rule: Each activity listed on one level is divided into two or more activities at the subsequent level. For illustration purposes, only one of the second-level activities—foundation—is expanded to the third level. The activities subordinate to the foundation activity are to *survey the building site, excavate and pour footers*, and *lay block foundation walls*. This increasingly detailed process would continue until the WBS reaches a point where the project can be effectively managed.

For the simple residential structure example in Figure 2.7, a three-level hierarchy is probably appropriate. At level 3, one of the activities for building the foundation, *survey the building site*, falls under the supervision of one surveyor or surveying crew (either internal or hired). Breaking this activity down further into a fourth level would tend to detail specific steps for the surveyor, something that would interfere with his or her work routine. For this particular hierarchical path in this simple WBS, a three-level hierarchy provides enough detail to define

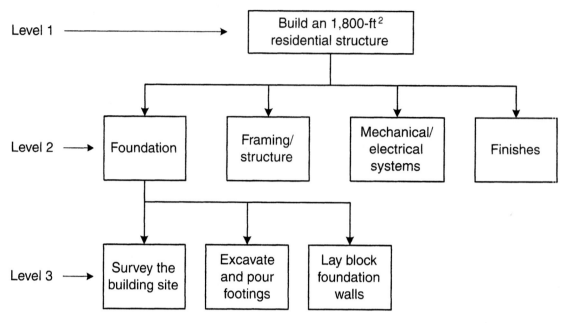

⟶ Build an 1,800-ft² residential structure

Level 2 ⟶ Foundation | Framing/structure | Mechanical/electrical systems | Finishes

Level 3 ⟶ Survey the building site | Excavate and pour footings | Lay block foundation walls

FIGURE 2.7　Residential structure WBS—levels 1, 2, and 3

responsibilities, estimate costs, and determine an activity schedule for the project. As the size and complexity of the construction project increases, so would the need for an increasing level of detail in the WBS.

The upper levels of the WBS tend to summarize large project components or provide decision **milestones** for management. Milestones are distinct events that provide crucial management information about the progress of the entire project or portions of it. Examples of milestones are the procurement of an important piece of equipment or the completion date of a critical project activity. Milestones are used to call attention to activities that are behind schedule to allow management to formulate a solution. The upper levels of the WBS are helpful to upper-level managers for the planning of costs and schedules of a project, especially in a multiproject environment.

Moving down through subordinate levels, or branches, of the reverse tree yields increasing detail of project activities and construction operations. The extent and nature of the work become more clearly defined as upper-level activities are subdivided into more detailed activities. At the upper levels, construction managers and planners are provided information about the project progress and about the assignment of accountability at subordinate levels of the WBS. At those lower levels, field supervisors and skilled labor are concerned about the responsibility for the specific steps to be performed to complete an activity and for the targeted completion dates. The WBS provides a means to identify responsibility for work activities and progression of work.

At some point, the number of levels or the activity detail defined in the WBS, called the **level of detail,** gets to a point where activities fall under a single worker, supervisor, subcontractor, or work division (e.g., electrical, mechanical, structural). The specific number of levels or the level of detail in a WBS is a matter of experience and practice for the planner. A small number of levels or the low level of detail may result in insufficient detail for effective planning, whereas too many levels or a too high level of detail may become unmanageable and increase planning costs. While there are no objective criteria to determine a specific level of detail in a WBS, there are a number of factors to be considered. As a general rule of thumb, the WBS has the appropriate level of detail when:

1. All elements of the WBS can be clearly communicated to the key players in the construction process.
2. Project progress relative to design, performance, and regulatory constraints can be observed and measured.
3. Accurate cost estimates can be developed.
4. Project activities can be scheduled with a definite start and completion date.
5. No significant time gaps exist between consecutive activities.
6. Project activities fall under a single worker or single supervisor and do not interfere with his or her work routine.
7. The required resources (i.e., materials, equipment, labor) for any particular activity do not significantly change over the life of the activity.
8. Reporting requirements for the project can be met.

As stated earlier, the WBS is a structured approach to enable project planning and scheduling, performance management and control, and cost analyses. Hendrickson and Au (1989) state that "it is generally the case that most schedules are prepared with too little detail than too much" (p. 276). Ultimately, the criteria used to determine an appropriate level of detail for the WBS depend on a combination of factors including the construction project's size and complexity as well as the planner's management and control needs.

2.3 EXAMPLE PROJECT—THREE-UNIT TOWNHOUSE

The purpose of this text is to introduce and thoroughly explain the concepts of and technical methods used for project planning and scheduling in the construction industry. Up to this point, the examples presented have been simple and direct (i.e., deck and small residential structure). The intent was to explain the concepts of construction planning and scheduling without having to explain much detail in complicated examples. Now that the construction project stages and work breakdown structure have been covered, it is time to move on to a more complex example. A detailed WBS and construction schedule are not normally prepared for most residential projects. However, for the purpose of this text, the example project is a three-unit townhouse that requires multiple activities to be completed, some occurring at the same time and others sequentially. It is complex enough to illustrate the methods used for realistic planning and scheduling, and yet not so large that the process becomes unmanageable for someone learning the fundamentals of project management. Most students using this text are also somewhat familiar with the construction process for this type of structure.

The details for the three-unit townhouse are provided in Appendix A, including layout drawings, project activity list/durations, work breakdown structure, and project schedule. These details will be referenced throughout the text depending upon the subject being discussed.

The three-unit townhouse is a two-story structure [Figure 2.8(a)] approximately 28 ft by 60 ft with two 1,125-ft^2 townhouse units (at approximately 18 ft by 28 ft per floor) and one 1,350-ft^2 unit (at approximately 24 ft by 28 ft). The layouts of the townhouses are relatively simple, with a living area, kitchen, and half bath (and extra bedroom for larger unit) on the first floor [Figure 2.8(b)] and two bedrooms (three for the larger unit) and full bath on the second floor [Figure 2.8(c)]. The structure is made of traditional materials using typical building techniques—wooden stud platform framing, sloped roof, and siding. Unlike a traditional residential house, firewalls are placed between the units, and other mechanical/electrical services are similarly located and shared where possible. Further, major building steps such as foundation and framing will be fabricated for the entire structure (all three townhouses) rather than separately or sequentially for single units.

The work breakdown structure for the townhouse structure is shown in Figure 2.9. The overall objective (level 1), *build a three-unit townhouse*, is broken down into seven major steps (level 2) including *prepare the site, build foundation, erect structure/frame, install mechanical systems, interior finish, exterior finish,* and

FIGURE 2.8(a)   Townhouse front elevation

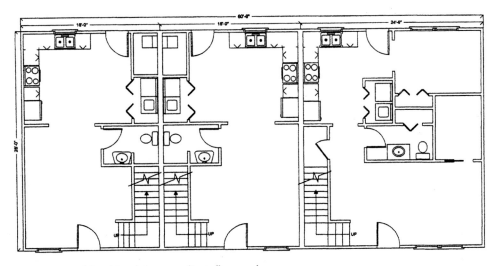

FIGURE 2.8(b)   Townhouse first-floor plan

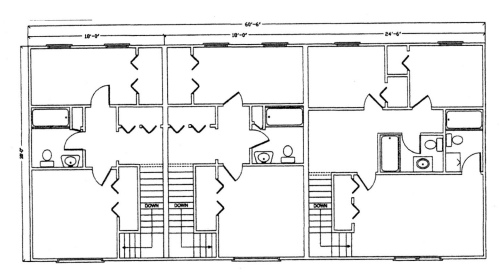

FIGURE 2.8(c)   Townhouse second-floor plan

inspection. These second-level steps of the WBS are further detailed into level 3. It is obvious that the activities given in level 3 of the WBS could be broken down into more detail. For example, the foundation activities of *excavate footings, place and cure concrete,* and *masonry* have multiple steps to complete. However, the WBS is not broken down to a fourth level because it is assumed that the previously stated criteria for an appropriate level of detail have been satisfied.

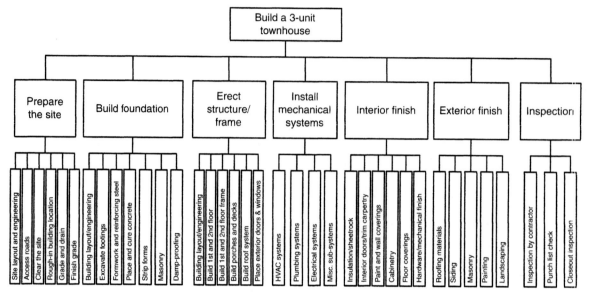

FIGURE 2.9   Townhouse WBS schematic

2.4   KEY PLAYERS IN THE CONSTRUCTION PROCESS

In Chapter 1, the construction industry and projects in construction were described. In this chapter, stages in the construction project have been explained. In each of these, the principal participants were mentioned and their individual roles partially explained in the context of the project life cycle (i.e., need, planning, design, estimation, construction, and occupancy/use). The intent in this section is to more fully explain the function of the key players and participants in the process and the organizational structures that facilitate their interaction.

❑ Owner
❑ Designer
❑ Contractor

2.4.1   OWNER

Most dictionaries include the terms *possess* and *control* in their definition of **owner**. The use of these terms to describe the owner is most appropriate in the context of construction projects. The owner *possesses* the construction project during the building process by funding the design and construction and after the building process by occupying and using the facility. The owner *controls* the construction project by developing the project concept or scope and by establishing basic objectives for its use (the facility's function or purpose) and appearance (the facility's form). The owner also *controls* or specifies, to some extent, the scheduled completion date of the project. Finally, the owner *controls* the overall design and construction cost of the project.

The owner initiates the construction project because of market demand or a perceived need for a facility or structure to be built. This need normally originates from one or more of these requirements: personal, commercial, industrial, or institutional. The owner can be either private or public—an individual, family, corporation, organization, or government enterprise that funds a project and ultimately uses the constructed facility. Public projects (e.g., schools, roads, airports, dams) are undertaken for the public good—to aid the citizenry. Public owners fund construction projects (see, for example, Figure 2.10) with public tax dollars and are typically government-oriented (i.e., local, state, or federal). Public projects require construction contractors to be chosen by the competitive bidding process. In this process, projects are publicly advertised and must meet certain performance, wage, and legal constraints.

Private owners have more flexibility in their selection of designers and contractors than do public owners because public tax dollars are not used. Privately owned construction projects also fall under less scrutiny than public projects although legal constraints such as zoning and building codes must be met for all applicable projects. A private owner may invite certain designers and contractors to bid on projects, especially when the owner has experience with particular companies. It is common practice for a private owner to forgo the competitive bidding process and handpick a contractor on the basis of the contractor's reputation and overall qualifications to do the job.

Whether public or private, the owner is the customer or client of the project designer and construction contractor. The owner sets the stage for project scope, budget, contract negotiations, and schedule of work. The owner oversees the performance of the project and approves the final product—the facility. The owner ultimately occupies and uses the facility and provides the designer, contractor, and the public with feedback, either negative or positive, on the quality of the designer and contractor. This eventually develops the reputation of the designer and contractor and will affect the future business of these companies.

2.4.2 DESIGNER

Most designers for construction projects can be categorized as either architects or engineers. An architect is a person who is skilled at the art of building, understands existing and future trends in architecture, and has the creative and technical ability to form plans and designs of structures. The architect has much to do with the appearance and function of a structure. An engineer has the ability to apply scientific and mathematical principles in a practical way to design and fabricate structures in an efficient and economical manner. The engineer has much to do with a facility's structural integrity, mechanical/electrical systems, and building infrastructure. Both the architect and engineer are professionally

FIGURE 2.10 Metropolitan construction site
SOURCE: *Construction Digest*

registered and/or licensed, with the primary purpose of ensuring public safety. Registration and licensing also provide the owner, as a customer, with some assurances about the quality of the design work.

It is common in construction projects for a design firm to have available both architects and engineers or those services. The design company works with the owner to develop the owner's general concepts into specific project objectives in terms of project scope, budget, and schedule of work. Many construction design companies further have the ability to negotiate and develop contract documents between the owner and the contractor.

Traditionally, the contractor does not begin the building process until the design from the architect/engineer is complete. This sequence is most common in the construction industry and is referred to as the **design-then-construct** (also referred to as the **design-bid-build**) process. An alternative to this process is the fast-tracking process. **Fast tracking** refers to the overlapping accomplishment of project design and construction. As the design of progressive phases of the project is finalized, these work packages are put under contract, a process commonly referred to as *phased construction*. Early phases of the project are commensurable under construction while later stages are still on the drawing boards.

2.4.3 CONTRACTOR

The conventional organization of the owner, designer, and contractor is as three separate participants. In other words, the owner works with the designer and contractor in different companies. The design is performed by the design company and the construction by the contractor. In this case, the contractor who has been awarded the construction project through either contract negotiations (private) or competitive bidding (public) is commonly referred to as either the **general contractor** or **prime contractor.** The prime contractor generally has a contract directly with the owner. Often the design company prepares these contract documents, or the designers assist the owner in development of the documents.

Prime or general contractors usually specialize in one type of construction such as residential, commercial, road/highway, earthmoving, or industrial. Prime contractors perform the majority or a portion of the work with their own crews depending on the particular project. When the work is the type with which the prime contractor is not experienced or not properly equipped, this work is usually contracted to subcontractors with special trade knowledge and skills, especially on large projects. Special trade subcontractors usually do one type of work, such as painting, carpentry, or electrical, or will do two or more closely related trades, such as plumbing and heating (see Figure 2.11). When a general contractor sublets a portion of the work to a subcontractor, the prime contractor remains responsible under his or her contract with the owner/designer for any negligent or faulty performance by the subcontractor. The prime contractor assumes complete responsibility to the owner/designer for the direction and accomplishment of the total work.

2.5   CONSTRUCTION PROJECT ORGANIZATIONAL STRUCTURES

There is no single organizational structure that encapsulates the relationships of the key players and their interactions throughout the life cycle of all projects. However, the owner-designer-contractor relationship is typically carried out through one of three types of organizational structures. One of the most common organizational structures used for construction projects, the traditional

FIGURE 2.11 HVAC subcontractor work on a commercial building
SOURCE: *Manus, Inc. (Lexington, KY)*

architect/engineer (A/E) structure, provides that the prime contractor has responsibility to the owner for the field construction only (Figure 2.12). This arrangement completely removes the contractor from the design process. Where the contractor provides construction services only, the usual arrangement is for a private design firm (architect/engineer) to perform the design in contract with the owner.

Two additional methods of organizing a construction project are the **design/build structure** and the **design/construction manager/adviser structure**. In the *design/build* organizational structure (Figure 2.13), the owner contracts with a single company that has the capabilities to both design and build a facility. This arrangement, also called the *design and construct* or *turnkey*, is generally used when a construction firm specializes in a particular type of construction and has available standard designs that it modifies to satisfy the owner's wishes. Usually the contractor has his or her own design section that includes architects and engineers as company employees. Basic to the design/construct process is the team concept. The owner, designer, and builder work cooperatively in the total development of the project. In some cases where time is critical, having the same organization both designing and building the facility allows *fast tracking* or construction to begin before completion of the final design.

With the *design/construction manager/adviser* structure (Figure 2.14), three separate firms are hired for design, construction, and construction management.

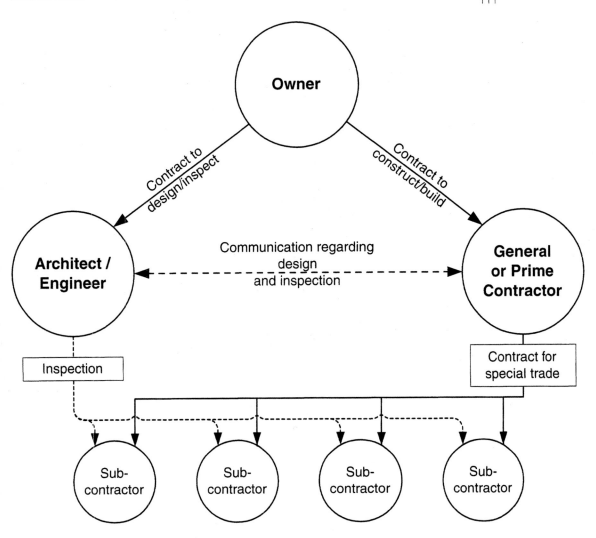

FIGURE 2.12 Key players in the construction process

The owner contracts directly with an architect/engineer for project design, a construction management firm for consulting, procurement, contract letting, and construction administration services, and a contractor for construction. In so doing, a design/construction team is created consisting of the owner, the designer, and the construction management firm (CM). The objective of this approach is to treat project planning, design, and construction as a system of integrated tasks and to communicate these tasks to all participants. During the design phase, the CM advises, with respect to performance criteria, construction feasibility of alternative building systems, site conditions, and availability of materials

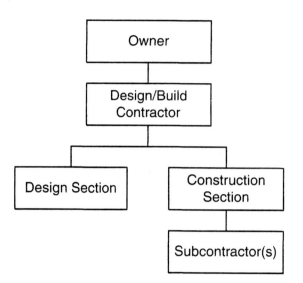

FIGURE 2.13   Design/build organizational structure

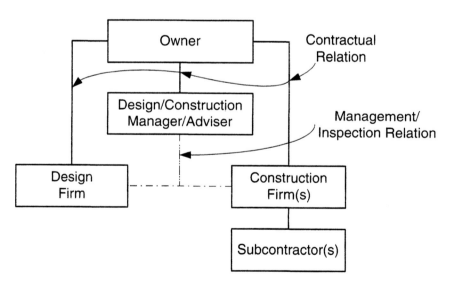

FIGURE 2.14   Design/construction manager/adviser organizational structure

and labor. During the construction operations, the CM assumes responsibility for the supervision, coordination, and administration of the project. This structure has been criticized because the CM typically has few economic incentives for optimizing the project schedule/resources and because the CM's objectives may come into conflict with those of the owner.

2.6 DESIGN-PHASE CONSTRUCTION MANAGEMENT

As construction projects become larger and more complex, it becomes more difficult for the owner and designer to assess the true cost and completion date of the project without the experienced opinion of the contractor. Larger projects also increase the need for larger construction contractors in terms of the number of employees and management capabilities. A construction management agency is sometimes hired as a consultant early in the design phase of the project to provide recommendations on construction techniques and operations, cost, and scheduling to the owner and designer.

When the construction project is very large, a project manager is sometimes hired to oversee the entire project for the owner. The project manager's position within the organizational structure of the overall project has broad authority and responsibility. In this situation, the project manager is likely to have authority over both the construction and design elements of the construction project. Both a construction manager and a design manager would manage their areas of responsibility, while the project manager would oversee these two positions and thus the entire project.

2.7 MANAGEMENT OF FIELD CONSTRUCTION

Houses, apartments, factories, offices, schools, roads, and bridges are but a few of the products of the construction industry. This industry's activities include doing work on new structures as well as making additions, alterations, and repairs to existing structures. Whether the process involves new or existing structures, the activities of construction projects must be organized, directed, and controlled by a supervisor. A myriad of titles exists for this supervisor— *constructor, construction superintendent, general superintendent, project engineer, project manager, construction manager, general construction manager,* or *executive construction manager.*

According to the *Occupational Outlook Handbook* (2000–2001) from the U.S. Bureau of Labor Statistics, "**construction managers** evaluate various construction methods and determine the most cost-effective plan and schedule. They determine the appropriate construction methods and schedule all required construction site activities into logical, specific steps, budgeting the time required to meet established deadlines. They are responsible for coordinating and managing people, materials, and equipment; budgets, schedules, and contracts; and the safety of employees and the general public." This description could be used to accurately describe a project manager as well.

In many construction companies, the construction manager and project manager may have very similar job descriptions and accountability. In fact, a project manager in one company may have the same duties, responsibilities, and authority as a construction manager in another. However, one important aspect of a project manager's position is that his or her duties are normally separate from those of field supervision. A foreman, site supervisor, or field superintendent handles the day-to-day direction of field operations. The project manager's duties involve working with the foremen, coordinating the subcontractors, directing construction operations, and keeping the work progressing smoothly and on schedule.

To be effective, the project manager must possess three essential attributes:

❑ Knowledge of the construction process
❑ Specialized technical management knowledge and skills
❑ People skills

A common characteristic one would find among project managers is an in-depth knowledge of the construction process, particularly technical knowledge. The project manager must have considerable background of practical construction experience so that he or she is thoroughly familiar with the workings and intricacies of the process.

The project manager must have, or have available to him or her, persons with expertise and experience in the application of specialized management techniques to the planning, scheduling, and control of the construction operations. Because much of the management system of a construction project is often computer-based, the project manager must have access to adequate computer support services.

In order to manage a technical process, the project manager must understand both the technology and the people involved. Knowledge of the technology is obtained through experience and/or education. People knowledge—interpersonal skill—comes mainly from experience with people in a management/business environment. While the technical background of a project manager is the basis of a thorough knowledge of the construction process, too much dependence on the technical aspect of management can create a conflict. Technical people are all about certainty, clarity, and understanding. However, the construction process, as with all other industrial processes that involve human interaction, cannot be entirely controlled with a good plan. Project managers, just like all technical managers, must think broadly, cope with uncertainty, and go beyond technology to embrace nontechnical solutions.

Because the project manager has overall responsibility for the project, he or she must have broad authority over all elements of the project. The nature of the construction process is such that the project manager often must take action quickly on initiative, and it is necessary that he or she be empowered to do so.

2.8 SUMMARY

Construction projects are done because an owner needs or desires a facility to be built. The owner has some conceptual ideas about the use and appearance of the facility and about a site or location for the facility. The owner contracts with a designer to develop project drawings and specifications and a contractor to procure materials, equipment, and labor and construct the facility. The contractor is selected by a competitive bidding process or negotiated contract.

The method widely used in the construction industry to break down the project into manageable components or work activities is called the work breakdown structure, or WBS. The WBS hierarchical structure can be thought of as an upside-down, or reverse, tree. The development of the WBS is an iterative process that considers the overall purpose of the project, design requirements, performance stipulations, and technical and logistical constraints. Moving down through subordinate levels of the reverse tree yields increasing detail of project activities and construction operations. The specific number of levels or the level of detail in a WBS is a matter of experience and practice for the planner based on factors such as communication needs, cost estimate accuracy, activity scheduling considerations, and supervision and reporting requirements.

The three principal players or participants in a construction project are the owner, designer, and contractor. The owner is either private (an individual or corporation) or public (a government organization). Public owners use public tax dollars to fund construction projects, and thus those projects fall under more technical and legal constraints than privately owned projects. The designer (architect/engineer) helps the owner develop general concepts into specific project objectives in terms of project scope, budget, and schedule of work. The prime or general contractor who is awarded the construction project through either contract negotiations (private) or competitive bidding (public) has a contract directly with the owner. Prime or general contractors specialize in one type of construction and perform the majority of the work with their own crews depending on the particular project. The prime contractor may also hire subcontractors with special trade knowledge and skills, especially on large projects. When a general contractor sublets a portion of the work to a subcontractor, the prime contractor remains responsible under the contract with the owner for any negligent or faulty performance by a subcontractor.

Three primary types of organizational structures are commonly used in construction projects: the traditional architect/engineer (A/E) structure, the design/build structure, and the design/construction manager structure. The A/E structure provides that the prime contractor has responsibility to the owner for the field construction only. In the design/build structure, the owner contracts with a single company that has the capabilities to both design and build a facility. With the design/construction manager structure, three separate firms are hired for design, construction, and construction management. The construction process is typically overseen by a project manager, who must have knowledge of the construction process, specialized technical management knowledge and skills, and interpersonal, or people, skills.

CHAPTER 2   QUESTIONS/PROBLEMS

1. Perform a search of technical and/or engineering journals to find an article with a construction management theme. The article should be at least five pages in length with a relatively current date (within the last five years). Using the article headings or major sections, break each section down into subtopics, assigning brief titles, and then create a title for each paragraph. Develop a work breakdown structure of the article with at least three levels of detail.

2. Perform an online search to find organizational structures of a large construction firm. Choose one, and describe the organizational structure in writing. Provide a printout of an organizational chart from that firm.

3. Compare and contrast the traditional architect/engineer and the design/ build organizational structures used to manage construction projects.

4. Describe the characteristics of an "effective" construction project manager.

5. Briefly describe the primary responsibilities of the key participants (i.e., owner, designer, contractor) in a construction project.

6. From the work breakdown structure for the three-unit townhouse in Appendix A, create a fourth level of detail from the existing third level. For each activity in the third level, generate at least two activities on the fourth level.

7. In the figure, the homeowner is proposing a wooden deck. The figure is not to scale, but critical dimensions are shown. The deck is designed for 2″ by 6″ framing on 16″ centers and 1″ by 6″ decking or flooring. The deck will be supported by nine 4″ by 4″ wooden posts that will be buried 18″

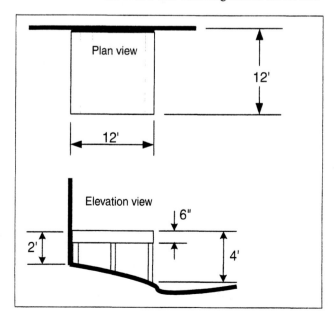

and surrounded by concrete. All activities will be manual labor with one carpenter and one laborer working. Assume all necessary construction tools are available. Develop a work breakdown structure of this job with three levels, identifying each task of the project. Schedule the tasks on an hourly basis by listing each task being started, continued, or completed per hour. Assume the deck will be completed, start to finish, in one 10-hour day.

8. *Group Activity*-Project planning and scheduling

   Student groups of four to six members should discuss and record answers to the question "Why is construction planning done?"

   - Assign an official recorder from your group to document group reasons.
   - Be sure that every individual in your group provides at least one reason to plan.
   - Spend 10 to 15 minutes discussing and recording answers.
   - Rejoin the class for a discussion on the different reasons that were proposed.
   - Ask someone to record the reasons on the board, overhead, or other group visual device.
   - Now define your group's project plans to the rest of the class and explain how the plans differ from project planning.
   - As a class, spend 3 to 5 minutes discussing the definitions and explanations that the groups presented.

REFERENCES

See References on page 199.

THE NETWORK PLAN

This chapter provides knowledge in the areas of:

- ❑ Definition of key terms
- ❑ Development of the network plan
- ❑ Development of the network model
- ❑ Project schedule diagramming techniques
- ❑ Critical path method

3.0   OVERVIEW

This chapter defines the key terms *project*, *activity*, and *milestone* with a detailed characterization of project activities. The chapter then explores the steps in the development of the network plan, including reviewing plans and setting objectives, analyzing the design for alternative construction methods, and developing the network model. The majority of the chapter explains the steps in developing the network model: defining, sequencing, and diagramming project activities, assigning durations, and calculating the schedule. Descriptions of four schedule diagramming techniques—bar charts, activity-on-arrow, activity-on-node, and time-scaled networks—are presented, along with advantages and disadvantages of each. Finally, the chapter discusses the critical path method, explaining that it is the basic methodology used widely to perform project scheduling for projects with particular characteristics.

3.1   DEFINITION OF KEY TERMS

You have learned about the unique nature of construction projects and the resulting nonstandardized processes used in construction. This uniqueness was also identified as a principal factor in determining the need to do project planning and scheduling in the construction industry. That need does exist, because without good planning and scheduling of project resources, the project will likely not accomplish its primary objectives of being completed on time and within budget. Before moving on to the development of the planning and scheduling process—in other words, the network plan—several key terms need to be defined.

PROJECT: We previously defined a **project** as a group of discernible tasks or activities that are conducted in a coordinated effort to accomplish one or more unique objectives. We can further state that a project is a set of jobs, tasks, or activities that must be completed in order to achieve specific project objectives, typically which are unique and nonrecurring. A project has several distinct characteristics. A project consists of well-defined tasks or activities that have a particular sequence or job order. Project activities may begin and end independently of each other within their sequence, and when all activities are completed, the project is completed.

ACTIVITY: A project **activity** is a discernible task or work function where a particular worker or crew of workers completes a specific item of work within a prescribed time frame. Project activities can be characterized as satisfying the following criteria:

1. *The activity outcome, work operation, or "job" must be measurable.* The activity must have a measurable objective that can be met as determined by the project participants. In most cases, a tangible result that can be clearly observed marks the end of an activity and is an appropriate measure of activity completion. However, the finish or end of an activity is not the activity itself as defined here but rather a milestone (defined below). The activity must also meet the criterion below of having a distinct start and finish time, not just the result after the activity is complete. A few examples include:

   - Excavating a footing for a residential structure foundation
   - Setting forms for a concrete retaining wall
   - Compacting the roadbed base along a specific section or length of highway (see Figure 3.1)
   - Performing the plumbing rough-ins on a townhouse

   These examples can be categorized as *production* or *construction* activities. These are the most common types of activities included in

FIGURE 3.1   Road base work
SOURCE: *Construction Digest*

the project schedule because they directly affect the duration or time outcome of the project. Other types of activities, procurement and management, also affect the schedule and must be considered for inclusion as well. *Procurement* activities are those where resources (i.e., materials, equipment, labor) are acquired or procured typically for production activities. *Management* activities are those administrative or support activities that affect the schedule outcome. Examples of management activities include approving and inspecting completed production activities, collecting material samples (e.g., concrete, steel) for lab testing, and collecting data for reporting purposes. For project planning and scheduling purposes, only those procurement and management activities that influence the project schedule should be included in the schedule.

2. *The activity has a distinct start and finish time.* The time required to perform an activity is called the *activity duration*, the time between its start and finish. The project planner determines the length of this duration based on the estimated materials to be used and crew production to complete the activity. Project activities occur in an ordered sequence; that is, an activity that immediately precedes another activity must be completed prior to the start of the second. The sequence determines placement of the activity within the entire project work and sets a distinct start and finish time in relation to the adjacent activities. Start and finish times of any activity can be recognized as project milestones although the finish is more

commonly the critical milestone. Completion of activities helps the project planner track the progress of the project.

3. *The activity must consume resources (i.e., time, material, equipment, labor, money).* Project activities consume resources. In other words, it takes time to complete the activity (i.e., duration, start-finish). The activity uses materials such as concrete, wood, and steel to fabricate or erect the facility being constructed. Normally workers are required from the activity start to finish, and these workers apply physical labor to the materials for fabrication or erection of the facility. In most cases, some form of tools or equipment is required to complete the activity to increase productivity or enhance safety, such as the crane placing structural framing steel shown in Figure 3.2.

4. *The activity must be assignable.* The project planner must be able to assign responsibility for project activities to a project team, supervisor, work crew, specialty trade, or project location. Activities must be defined to the level of detail that clearly assigns the activity to a supervisory team or person. This criterion is critical for good communication during project planning and scheduling as well as during the actual construction process. If a contractor's own work crew is assigned a particular activity, the project manager or field supervisor of that crew is given that responsibility. If work is to be done by a specialty contractor or subcontractor, the responsibility is assigned to that supervisor.

MILESTONE: The start or more likely the finish of a project activity identifies the previously mentioned project **milestone.** Milestones are key events in the life of a project and provide the project planner points in time to measure project progress or phases of project completion. Unlike a project activity, the milestone does not have a distinct start and finish time, nor does it consume project resources. A few examples of a milestone are beginning excavation, getting a house under roof, completing the foundation, and putting structural steel in place.

3.2   DEVELOPMENT OF THE NETWORK PLAN

Management of the construction process must be performed regardless of project type or size. Construction activities have to be defined and ordered, and resources have to be allocated to complete the work. Many simple or small projects are managed with minimal formal planning of work activities. However, you read in Chapter 1 that the need for construction planning and scheduling increases as project proportions increase. For the purpose of discussion in this chapter, the assumption is made that for all references to construction projects, formal planning and scheduling of work activities is required due to project size and complexity.

FIGURE 3.2 Crane placing framing steel
SOURCE: *Construction Digest*

A construction project presents the contractor with a difficult and challenging problem. The contractor commences the project with a facility design (i.e., technical drawings and specifications), which provides a very detailed "picture" of the end product or completed facility. The problem exists because the product or facility is represented as finished or complete and few details are provided the contractor about how he or she must perform the construction activities to complete the work. The contractor must use the technical descriptions of the completed facility to form a plan that executes all aspects of the work to physically construct the facility. The plan must contain all required resources (i.e., time, materials, equipment, and labor) and must be such that the facility is completed on time and within the estimated budget. The process used by the contractor to formulate a construction plan is called the **network plan.**

Fundamental to the construction planning and scheduling process is the development of a network plan. The term *network* is used here to identify the construction planning process as a system of connected activities functioning to accomplish project objectives. Further, it represents complex groups of communication channels that cross or interconnect as a system. The network plan involves reviewing the owner's conceptual plan and setting objectives, analyzing the design for alternative construction methods, and developing the network model. Steps in the development of the network model include defining work tasks, setting their sequence, diagramming the network, assigning activity durations and required resources, and calculating the schedule. The steps to develop the network plan are to:

❑ Review the plan and set the objectives.
❑ Analyze the design for construction methods.
❑ Develop the network model.

3.2.1   REVIEW THE PLAN AND SET THE OBJECTIVES

The contractor carefully examines the design documents and communicates with the owner and designer to fully understand the owner's conceptual idea for the facility in terms of function and form (use and appearance). Broad objectives are established for the project start and completion dates, rate of progress for major project phases, budget, and expected completed facility. Interim milestones are established, and responsibility for the primary project phases is assigned.

3.2.2   ANALYZE THE DESIGN FOR CONSTRUCTION METHODS

Based on the owner's conceptual plans/ideas, the designer generates a detailed technical description of the facility. During this process, the designer must consider many design alternatives to fulfill the owner's need for facility use and appearance. Similarly, the contractor is faced with the prospect of carrying out various activities in the project using alternative methods. The level of design uniqueness and complexity can intensify this problem. While there are basic standard methods for some construction work (e.g., framing a residential structure), alternative methods are available to the contractor to complete many construction activities. For example, concrete placement in an elevated location could be accomplished using either a combination of crane and bucket or an auger pump system (see Figure 3.3). For each system, decision factors such as cost, equipment availability, and worker/supervisor experience with the technology must be considered.

FIGURE 3.3   Gantry crane and concrete bucket
SOURCE: *Construction Digest*

According to a recent survey of 240 project management professionals in the construction industry, as construction projects get more complex, computer software usage for project management purposes increases (Liberatore, Pollack-Johnson, and Smith, 2001, p. 103). Adding to this complexity are the varied analyses that the project planner must perform in determining the most effective method to use to complete activities in a construction project. Computer use for construction project planning and scheduling gives the project planner the ability to perform "what-if" analyses on these various alternatives, enhancing the planner's decision-making capabilities.

3.2.3 DEVELOP THE NETWORK MODEL

The network model is the tangible result of the network plan used for management purposes. The network model, as used in construction planning and scheduling, is a schematic representation of the construction process that describes known, estimated, and/or calculated properties and may be used for further analysis of its characteristics. An accurate model of the construction process is critical for the effective management of the project. However, the model is only a representation of the actual construction process and performs only as well as the data used for its generation. A model developed with inaccurate data not only misrepresents the process but also may compromise the construction plan. The saying "garbage in—garbage out" holds true here. The basic steps in building the network model are as follows:

- ❑ Define work tasks or activities.
- ❑ Put activities in order or sequence.
- ❑ Diagram the activity sequence in network format.
- ❑ Determine activity durations.
- ❑ Calculate the schedule and adjust it to meet constraints.

3.2.3(a) Define work tasks or activities

Defining project activities provides the framework for scheduling the construction work, allocating the necessary resources for activities, and setting the sequence or order for activities. The defined activities must represent all necessary work tasks within the project and include all elements of the design. The work breakdown structure (WBS) is a very practical and useful method to define project activities. The WBS provides an organized method to break down or divide a construction project into manageable parts using a hierarchical structure that looks like an upside-down tree. This "reverse-tree" method defines the project in increasing levels of detail by dividing project objectives into subordinate-level activities. This process continues until the project activities or specific work tasks meet several general guidelines—the activity is under a single supervisor and single unique crew of workers; it is performed continuously or through a continual time segment; it has a defined start and stop time. A more comprehensive list of criteria for determining an appropriate level of detail for the WBS is provided in Chapter 2. It is important to remember the intent of defining activities and generating a schedule. The construction plan and schedule is a management tool to be used by various levels of project planners and supervisors; it is not simply a detailed description of the project and the construction process.

The project planner is best served by relying on field supervisors and, if possible, the people who actually perform the work to define activities as well as estimate durations and allocate necessary resources. The planner needs to call upon those persons with the most experience and knowledge about the actual construction work. It may be helpful to use the following guidelines to categorize activities. They may be defined by:

1. Responsibility—specific supervisor, work crew, or subcontractor
2. Type of work—general (e.g., framing, excavation) or specialty (e.g., plumbing, electrical, HVAC)
3. Type of equipment used (e.g., crane, backhoe, auger)
4. Type of materials used (e.g., wood, concrete, steel)
5. Type of structural elements (e.g., frame, foundation footings, walls)
6. The owner's breakdown of the work for bidding and payment purposes
7. The contractor's breakdown of the work for estimating and cost accounting purposes

3.2.3(b)   Put activities in order or sequence

The order or sequence of any activity within a project depends upon where (or when) it occurs in the overall sequence of events to complete the project. Not all project activities can start at the beginning of the project. Of most importance is the identification of those activities that must immediately precede an activity (immediate predecessors) and those that immediately succeed an activity (immediate successors). The sequence of defined activities or order of events in a construction project is called the activity *precedence* within the project. Because we have referred to the modeling of the planning and scheduling process as a network (the *network model*), the project schedule can be referred to as the *precedence network* or ordered system of events. It is not uncommon to represent the sequence or order of project activities by listing activities with accompanying immediate predecessors (and sometimes successors), as shown in Table 3.1.

At this point in the planning and scheduling process, it is best to consider time as the only resource constraining project activities. Other resources (i.e., materials, equipment, labor) may be considered at a later time once the time schedule of activities has been established and other appropriate resource requirements for activities have been identified. The consideration of time as the only constraint at the beginning of this process compels the project planner to mainly consider the *physical constraints* on the activities. Physical constraints are those

TABLE 3.1 Example of a Building Foundation Predecessor List

ACTIVITY	DESCRIPTION	IMMEDIATE PREDECESSOR
A	Clear site and deliver material	—
B	Excavate footings	A
C	Pour concrete footings	B, D, E
D	Cut and bend reinforcing steel	A
E	Set and tie reinforcing steel	B, D

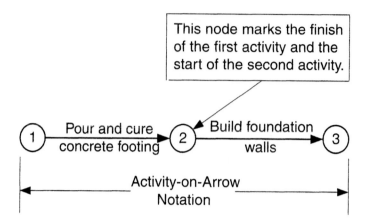

FIGURE 3.4   Activity sequence resulting from physical constraint

that exist due to the physical process of construction, where one activity must be physically complete prior to the start of another activity. As an example, the concrete footings of a structure must be poured and the concrete cured prior to building the foundation walls (see Figure 3.4). As other required resources are identified for project activities, these resources may constrain the project schedule as well. An example of a *resource constraint* is the conflict that may occur when two activities scheduled at the same time (in parallel) during a project each require the same piece of equipment, such as a large crane. If two cranes are available, the two activities may proceed in parallel and no scheduling conflict exists. Otherwise, one of the project activities must be delayed until the crane is released from the first activity. Other types of constraints, including safety, environmental, and management, are further detailed in Chapter 6.

3.2.3(c)   Diagram the activity sequence in network format

You have learned that the work breakdown structure is a process of dividing upper-level objectives/activities into lower-level activities within specific work groups (Chapter 2). The groups represent the major functions or phases of the construction process. As an example, the activities defined to construct a building foundation as listed in Table 3.1 are subordinate to the foundation activity/objective. To further illustrate this, note that the three-unit townhouse in Appendix A lists seven activities subordinate to *Build foundation* and six subordinate to *Erect structure/frame*.

The work groups and subordinate activities produced in the WBS are a good starting point to develop schedule diagrams for the project. Often the project planner will begin the diagramming process by establishing a sequence of activities within groups. This method maintains relatively small, manageable sets of activities and allows the project planner to assign supervisor, specialty, and/or department responsibility for the work. The small sets of activities are sequenced

Activities and nodes in Figure 3.5 are identified by letters (activities A, B, C, . . .) and numbers (nodes 1, 2, 3, . . .). In Figure 3.6, activities are identified by letters (activities A, B, C, . . .). In these figures and for the further explanation of schedule diagramming techniques later in this chapter, the letters and numbers do not specifically represent defined project activities but rather simply ambiguous activities. This gives you the opportunity to concentrate on the concepts of project scheduling rather than individual descriptions of activities. More realistic examples are provided in detail in subsequent chapters.

or ordered with other sets into larger and larger groups until the entire project has been included in the schedule.

Project activities are represented graphically in two primary types of diagrams—the bar chart and the sequence diagram. Simply put, the bar chart shows the timing of project activities within the entire project duration, whereas the sequence diagram shows the sequence or order of activities relative to one another. The bar chart, commonly used in business and education, represents data in horizontal or vertical, rectangular bars whose lengths are proportional to activity durations (Figure 3.5). The sequence diagram represents activities and their relationships through a system of nodes and arrows. There are mainly two types of sequence diagrams—the activity-on-arrow, or AOA, and activity-on-node, or AON (Figure 3.6). While the bar chart and sequence diagrams are most commonly used for scheduling in the construction industry, other types of scheduling diagrams are available; one of the most notable is the time-scaled network. Further details of the bar chart, sequence diagrams, and time-scaled network are provided in the next section of this chapter.

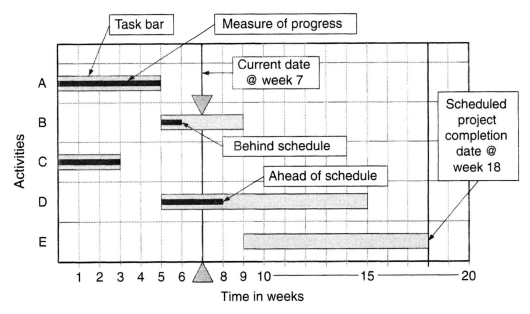

FIGURE 3.5 Bar chart schedule

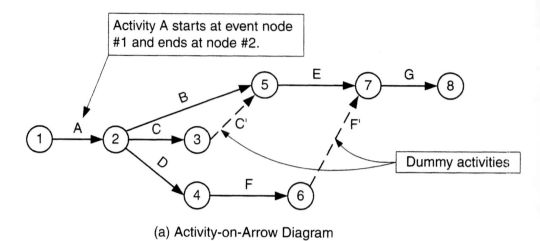

Activity A starts at event node #1 and ends at node #2.

Dummy activities

(a) Activity-on-Arrow Diagram

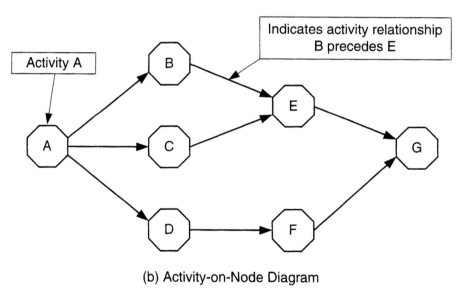

Activity A

Indicates activity relationship B precedes E

(b) Activity-on-Node Diagram

FIGURE 3.6   Sequence diagrams

3.2.3(d)   Determine activity durations

According to our definition of a project activity, the activity must consume the resource of time and have a distinct beginning and ending time (Section 3.1). The *activity duration* is the time it takes to complete the activity from its start to finish. This time can be represented in minutes, hours, days, weeks, or other units. The time unit of *days* is often used in construction scheduling because many construction activities require multiple days (two or more) to complete. Use of the unit of *hours* tends to make the schedule too detailed, and the unit of *weeks* is not detailed enough. The time unit of days is a general rule of thumb and should not be considered absolute. The project planner must decide the appropriate time unit, but once that decision is made, consistent time units should be applied to all activities in the schedule. The use of computer software for project scheduling allows much flexibility in assigning time units. The units can be modified for the entire schedule (e.g., changed from hours to days) to reflect the level of detail required for effective communication to the appropriate responsible party. For example, if a project manager is communicating the progress of several projects to upper-level management, the software can be adjusted to reflect the time in weeks for a particular project. That same project manager may adjust the project to time units of days or even hours to communicate responsibility for certain project activities to a field supervisor.

The project planner determines the length of each activity duration based on the estimated materials to be used and crew production to complete the activity. In lieu of calculating each activity duration for each new project, the planner may also use historical data from previous projects completed by the contractor. If historical records are available for particular types of activities that may be repeated in a project being planned, an average time for those certain activities can be determined. However, the unique nature of construction activities due to variations in facility designs and job site environmental conditions requires caution in the use of and particularly dependence on historical data. Some basic guidelines for estimating or determining activity durations during the planning and scheduling stages are as follows:

1. When using historical data, carefully check the reliability of the data and be certain that the work of past activities precisely matches that of the activity in question.
2. Estimate/determine the duration of activities one at a time. Do not consider other preceding, succeeding, or concurrent activities when making this determination.
3. For a given activity, assume that materials, labor, equipment, and other resources are available. Do not consider resource requirements of other preceding, succeeding, or concurrent activities.
4. Be aware that for each activity the level or amount of labor and/or equipment has been determined from the contractor's cost estimate. In that estimate, often a normal amount of labor and/or equipment is assumed, using conventional crew sizes and equipment allocations. During the duration-estimation process, ignore conflicting demands among concurrent activities for resources.
5. Use a normal, defined work schedule. Do not consider overtime or multiple shifts unless this is a usual procedure or a part of the standard workday.
6. Concentrate on determining the duration of the individual activity. For the purpose of activity-duration estimates, temporarily disregard

all other time considerations. In particular, the planned or desired completion date of the project should be entirely out of mind.

7. Use consistent time units for all activities.

Let's start by applying a function for material use and worker productivity rates to determine accurate activity durations. The basic function to calculate the duration, D_i, of activity i is

$$D_i = Q_i/P_i \tag{3.1}$$

where Q_i is the quantity of work to be performed and P_i is the productivity of the work crew(s) performing the work. The quantity of work, Q_i, depends on the type of work required for the activity. It is expressed by the type of material used for the activity, such as the cubic yards of concrete to be placed or earth to be excavated, the square yards of concrete formwork to be built, the linear feet of electrical wiring to be installed, or the number of structural steel members (e.g., beams, girders, columns) to be assembled. The productivity of the work crew(s), P_i, is expressed as the amount of work that can be completed for the specific material type and its associated unit of work.

For example, suppose a 1-cubic-yard (or 1 yd^3; or 0.76 cubic meter, or 1 m^3) hydraulic excavator (see Figure 3.7) can excavate about 720 yd^3 (550 m^3) of ordinary earth—sandy clay/moist loam—during a normal 8-hour workday. If 1,700 yd^3 (1,300 m^3) needs to be moved, it would take approximately 2.4 days to complete the work, or

$$D_i = \frac{1,700 \text{ yd}^3}{720 \text{ yd}^3/\text{day}} = 2.4 \text{ days}$$

To find the quantity of work in Equation 3.1 for all construction activities, a detailed analysis of the facility design must be conducted. Design documents (i.e., drawings and specifications) are carefully examined to determine the materials needed for project activities. These *quantity takeoffs* specify the types and quantities (e.g., area, volume, linear feet) of materials needed for an activity. Quantity takeoffs are necessary for accurate estimates of project costs, whether the project is contracted through negotiation or competitive bidding. Once the types and quantities of materials for an activity have been determined, the construction method must be decided upon in order to determine the specific work to be done. A general or specialty crew of laborers is identified, and the number of workers is assumed so that their productivity rate can be estimated. Average productivity rates for standard activities can be found in several sources, including *Walker's Building Estimator's Reference Book* and *Means Estimating Handbook* (see the references list at the end of this chapter for complete citations).

The quantity of work, Q_i in Equation 3.1, is a result of quantity takeoffs performed by an estimator(s) or project planner(s). Assuming reasonable accuracy of the takeoffs, the resulting quantities are conclusive with little or no variation. They

FIGURE 3.7 Hydraulic excavator loading an articulated truck
SOURCE: *Construction Digest*

are said to be *deterministic*. This cannot be said for the work crew productivity, P_i in Equation 3.1. Worker or work crew productivity is said to be *stochastic*. This means that productivity varies infinitely between some maximum and minimum, and an average value must be determined based on past crew performance, complete activity information, and good judgment. Variation in productivity is based on factors such as worker experience and attitude, job site environmental conditions, and crew interaction/interference with other crews performing other activities.

In the example above, the excavator digs about 720 yd^3 (550 m^3) of ordinary earth during an 8-hour workday. Normal work conditions are assumed for this level of productivity—the operator has at least average skill level, the

It is not unusual to switch the order of the steps in Sections 3.2.3(c), "Diagram activity sequence in network format," and 3.2.3(d). "Determine activity durations," in this process. Once the activities are defined, sometimes the planner will first assign activity durations to each activity and then produce the sequence diagrams.

equipment does not have unscheduled downtime, and the work site has optimum environmental conditions. If the operator, equipment, or environmental conditions worsen (e.g., the soil is very wet or is frozen), then this productivity value will likely be less than 720 yd^3. The difficulty that arises is accurately estimating this reduction. The adjusted productivity must be based on sound judgment and, if available, on the expert opinion of someone knowledgeable with the factors that effect work conditions and more specifically the magnitude of that effect on productivity. It's not simply a matter of understanding that productivity will drop; the planner must assign a specific, lesser value to that productivity. Regardless of the accuracy of activity durations during the planning and scheduling stages, it is possible that the work conditions may change during the project. This requires careful monitoring and ongoing adjustments in the schedule from the beginning of the planning stage to the end of construction.

THREE-UNIT TOWNHOUSE: The durations for activities for the three-unit townhouse are provided in the activity code table in Appendix A. The durations, in days, were determined using normal crew sizes as indicated working four 10-hour workdays, a common technique in the construction industry. Productivity rates are based on standard values from the *Means Estimating Handbook* (Mahoney, 1990).

3.2.3(e)   Calculate the schedule and adjust to meet constraints

Let's consider the information that we have up to this point. We've defined project activities, put them in sequence, diagrammed the network, and determined and assigned durations to each activity. We can now use these data to calculate the schedule and make necessary adjustments. Calculating the schedule involves setting a beginning date for the project and numerically determining the start and finish times of individual project activities, relative to their sequence and durations. The actual mathematical calculations performed in a schedule are almost exclusively done using some type of scheduling software such as Primavera Project Planner (Primavera Systems, Inc., Bala Cynwyd, Pennsylvania; see Feigenbaum, 1998).

The scheduling process determines the specific sequence or path of activities— the *critical path*—that must occur without delay for the project to be completed at its earliest time. The critical path is found by examining the earliest and latest start times and earliest and latest finish times for all activities. Those activities that have equal early and late start (and thus equal early and late finish) times are critical—they have no flexibility in their start (and thus finish) time. The activities that have a difference in their early and late start and finish times have flexibility, or "float" time, and are said to be *noncritical*. Management of the critical and noncritical activities and available float time is what makes scheduling a valuable tool for the project planner. If constraints are placed on the schedule, such as a definite finish or "need this done by" date, the planner can then make adjustments in the schedule to meet these constraints. The specific analysis techniques used in scheduling need to be known to fully understand the concepts of scheduling, and they are explained in detail in subsequent chapters.

3.3   PROJECT SCHEDULE DIAGRAMMING TECHNIQUES

Several methods are available to the project planner to graphically represent project schedules. Each has a specific purpose generally distinguished by the type of information being communicated. Four methods are described here including the bar chart, two sequence diagrams (activity-on-arrow and activity-on-node), and the time-scaled network. While these four methods are well known in the construction industry, the bar chart and activity-on-node diagrams (i.e., precedence diagramming method) are the most widely used and the two methods most heavily emphasized in this book.

3.3.1   BAR CHART

We have stated that the bar chart, commonly used in business and education, represents data in horizontal or vertical bars whose lengths are proportional to activity durations (Figure 3.5). Bar charts used for scheduling activities and workers date back to the early 1900s. Polish scientist Karol Adamiecki (1931) developed the "harmonygraph," a bar chart—type graph with a time scale represented on the vertical axis and activities in columns along the horizontal axis. However, most credit for development of the bar chart for project scheduling goes to the shipbuilder Henry Gantt during World War I. Gantt developed what is called the *Gantt chart*, which is a horizontal bar chart used for scheduling multiple overlapping tasks over a time period.

Bar charts are widely used today for project scheduling because they are relatively easy to construct and are the best diagramming technique to visually represent activity progress in the project. As can be seen in Figure 3.5, activities are listed on the vertical axis and are presented as task bars whose horizontal placement depicts the start and finish times for activities. For example, activity A in Figure 3.5 begins at the start date of the project and ends at the end of week 5 in the project for a total activity duration of 5 weeks. The current date, week 7, is shown as a vertical line on the chart. Here the darkened, smaller progress bars within the task bars indicate that activity B is behind schedule and activity D is ahead of schedule. The only activity that has not started is activity E, and the scheduled completion date of the project is the end of week 18.

Advantages of Bar Charts

❑ Simple and quick to generate.

❑ Easy to use and interpret.

❑ Great communication tool at all organizational levels.

❑ Project plan and schedule can be shown together.

❑ Project progress can be represented graphically.

Disadvantages of Bar Charts

❑ Planning and scheduling are considered simultaneously.

❑ Not easily modified.

❑ Do not show activity dependency relationships (sequence).

❑ Awkward for large projects.

3.3.2 ACTIVITY-ON-ARROW DIAGRAM

In the AOA diagram, activities are represented by *arrows* and the start and finish of those activities by *nodes*. For example, activity A in Figure 3.6(a) is defined at its start by node 1 and its finish by node 2. Node 2 not only designates the completion of activity A, but also is the beginning of activities B, C, and D. The use of nodes to show the start and finish of project activities can be traced back to the initial development of PERT and CPM. Both the PERT and CPM techniques required projects to be represented graphically. In the first publications on the critical path method, the activity-on-arrow method was used. The method caught on and was used for years to represent projects. The concurrent development and use of high-speed computers to solve scheduling networks reinforced the use of the AOA methods. Many computer programs written for solving CPM and PERT problems used the predecessor-successor (node 1 to 2) nomenclature.

The AOA and AON diagramming techniques were a result of two significant developments in the field of project management in the 1950s—PERT and CPM. The PERT, or program evaluation and review technique, was developed to aid in producing the U.S. Polaris missile system in record time in 1958. The PERT procedure provides a probability that a project will be completed on or before a specified completion date based on variable time estimates of activity durations. The critical path method, or CPM, grew out of a joint effort in the late 1950s by the E. I. duPont de Nemours Company and Remington Rand (Moder, Phillips, and Davis 1983, 13). CPM was developed to determine how to reduce project costs by completing the project in the minimum time possible.

A unique feature of the AOA diagram is the required use of *dummy activities* to represent activity relationships. Because arrows represent activities and because nodes tie one activity to another or designate immediate predecessor-successor relationships, the need arises to use *dummy activities* to show multiple activity relationships. For example, activity E in Figure 3.6(a) is preceded by both activities B and C. However, activities B and C must finish on separate nodes—3 and 5, respectively. Therefore, C′ (pronounced "C prime") must be inserted between the end of activity C (node 3) and the beginning of activity E (node 5). As you can imagine, this process becomes very complex for large projects. It is for this reason that AON diagrams are more widely applied in most scheduling software programs and used by professionals in the construction industry. A more detailed explanation of both the AOA and AON diagramming techniques is given in Chapter 4.

Advantages of AOA

❏ Oldest sequence diagramming technique.
❏ Easy to interpret activity start and finish times as separate events.
❏ Good communication tool for showing activity sequence.

Disadvantages of AOA

❏ Use of dummy activities is cumbersome and confusing.
❏ Time is not shown graphically.
❏ Does not easily show project progress.
❏ Awkward for large, complex projects.

3.3.3 ACTIVITY-ON-NODE DIAGRAM

The AON diagram represents activities as nodes, and the sequence of activities is shown with arrows, as can be seen in Figure 3.6(b). The process for AON diagramming requires only the activity names and their sequencing in the project relative to other activities. For any particular activity, the sequence is designated by those activities immediately preceding and immediately succeeding the activity in question. These immediate predecessors or successors are shown with a line and arrow joining two activity nodes. For example, activity A in Figure 3.6(b) immediately precedes activities B, C, and D, and these three activities are the immediate successors of activity A. Upon examination of Figure 3.6(b), it could be said that activity A also precedes activities E, F, and G, but A is not their immediate predecessor. The immediate predecessor and successor list for Figure 3.6 is given in Table 3.2.

TABLE 3.2 Predecessor List for Figure 3.6

ACTIVITY	IMMEDIATE PREDECESSOR	IMMEDIATE SUCCESSOR
A	—	B, C, D
B	A	E
C	A	E
D	A	F
E	B, C	G
F	D	G
G	E, F	—

Advantages of AON

❏ Eliminates the need for dummy activities, which are needed in the AOA diagram.

❏ Good communication tool for showing activity sequence.

❏ Easy to use and interpret—easy to learn.

❏ Great communication tool at all organizational levels.

Disadvantages of AON

❏ Time is not shown graphically.

❏ Does not easily show project progress.

❏ Is not easily converted to an AOA equivalent for those wishing to use that technique.

3.3.4   TIME-SCALED NETWORK

A time-scaled network diagram combines the principal features of the bar chart and the sequence diagram. This technique graphically represents time as in the bar chart and the activity sequence as in the AOA diagram. The project is plotted on a horizontal time scale, with arrow vectors and nodes representing activities and with arrow lengths representing time, as shown in Figure 3.8(a). The technique displays the logical connection between activities in the context of a time scale in which each horizontal position represents a point in time. The time-scaled network can also be drawn as a bar chart with network logic, as in Figure 3.8(b). Most scheduling software programs allow the user the ability to add (or "turn on") relationship arrows to bar charts. While this technique is very useful to show both activity time and sequence as well as represent project progress, the resulting diagrams are so crowded with information that communication effectiveness for most projects is impeded.

Advantages of Time-Scaled Network

❏ Shows activity sequence or order.

❏ Project plan and schedule can be shown together.

❏ Project progress can be represented graphically.

Disadvantages of Time-Scaled Network

❏ When in AOA mode, use of dummy activities is cumbersome and confusing.

❏ Not easily modified.

❏ Awkward for large, complex projects, reducing effective communication.

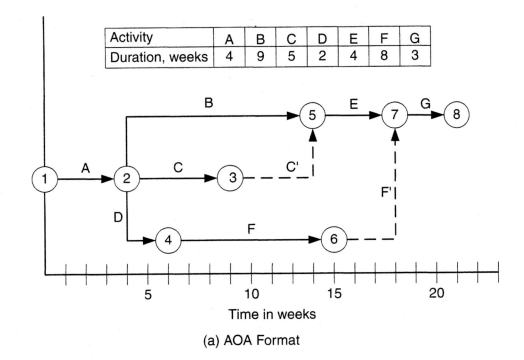

Activity	A	B	C	D	E	F	G
Duration, weeks	4	9	5	2	4	8	3

Time in weeks

(a) AOA Format

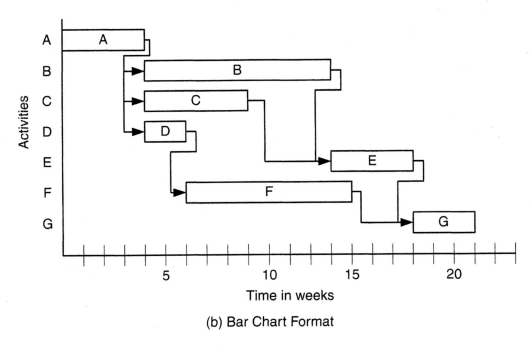

Time in weeks

(b) Bar Chart Format

FIGURE 3.8 Time-scaled network

3.4   CRITICAL PATH METHOD

The critical path method was developed in the late 1950s to aid in the planning and scheduling of large projects. Created by a joint effort of the E. I. duPont de Nemours Company and Remington Rand Corporation, CPM was used to determine how to reduce the time required to perform routine plant overhaul, maintenance, and construction work. The intent was to reduce project costs by finding the "optimum tradeoff of time (project duration) and total project cost" (Moder et al., 13). CPM was quickly recognized by professionals in government and industry as an effective cost-saving tool. In fact, during its early years, U.S. government agencies encouraged the use of CPM by contractors to schedule major government projects (Wiest and Levy 1977, 1). Today, the critical path method is widely used for project scheduling to communicate project details at many levels of management. The terminology of CPM—activities, predecessors, successors, durations—is well known by project managers, engineers, and field supervisors. Most project scheduling computer software uses the CPM as the basis for solution algorithms.

Basically, the critical path method prescribes the project modeling process described in Section 3.2.3 of this chapter, where activities are defined, sequenced, and diagrammed and where durations are assigned and the schedule is calculated. The mathematical analysis in CPM is performed using the data of activity ordering (predecessor-successor) and durations. The process results in the minimum completion time for a project as well as the start and finish times of each activity. The *critical path* represents the sequence or path of activities that takes the longest to complete, and all activities along this critical path are termed *critical activities*. The length of the critical path or sum of all critical activity durations is equal to the minimum project duration. Any delay for a critical activity means the project duration is increased by the amount of that delay. All activities not on the critical path are *noncritical* and have some level or amount of flexibility or float.

To properly apply the CPM, several assumptions are made regarding projects and the use of CPM for computing schedules, as follows:

❑ The project can be divided into well-defined tasks or activities, which when completed mark the end of the project.
❑ Project activities are ordered or follow a specific sequence of work to complete.
❑ An activity that immediately precedes a second activity must be completed prior to the start of second activity. An important extension of this assumption, called precedence scheduling, is covered in Chapter 5.
❑ Project activities, once defined, may be started and stopped independently of each other, within a given sequence.
❑ The amount of time to complete a project activity is known with reasonable certainty, or the activity duration is *deterministic* or not

highly variable. Those projects where variable time durations apply to activities are explained with the PERT process in Chapter 8.

❑ Contractors have some level of experience managing similar projects and estimating/determining project activity sequence and durations.

❑ Scheduling calculations are, at least at first, based on fixed activity durations and well-defined predecessor relationships. No resource constraints are recognized other than those imposed by activity time and sequence.

The critical path method is widely used in business and industry on projects that meet the assumptions stated here, especially the construction industry. Moder et al. (1983) pointed out two decades ago that "construction projects tend to be the largest individual area in which these methods are applied" (p. 18). This still holds true today because CPM allows project planners and other participants to evaluate alternative construction methods and allocation of resources in the planning stage rather than in the field after construction has begun. The following advantages of CPM can be realized in the planning and scheduling of construction projects:

1. *Organization.* The critical path method requires contractor management to set objectives and carefully plan and organize time and resources for project work. This occurs during the planning stage, which allows the contractor to identify potential problems early and plan for their implementation. The contractor can concentrate much effort on those activities that are the most critical and constraining in the schedule.

2. *Communication.* Important products of the critical path method are graphical diagrams of the project, which show activity timing and/or sequence as well as project duration and progress. These diagrams can be used to effectively communicate project plans and implementation to all project participants. Further, the diagramming process tends to bring together personnel involved in the project whose combined efforts result in a project and resource schedule. This encourages involvement and develops a sense of working together toward a common goal.

3. *Utilization.* As the techniques of the critical path method have progressed since the late 1950s, so has the development of personal computer hardware and software for project scheduling. Today, a variety of computer software based on CPM is available to project planners to manage most projects regardless of size and complexity. Additionally, university academic programs teach CPM as the main scheduling technique and commonly include computer software in project scheduling classes. Thus, many project planners have exposure to CPM and CPM-based software prior to taking a management position in industry.

3.5   SUMMARY

To fully understand the techniques of project planning and scheduling, it's necessary to understand the key terms *project*, *activity*, and *milestone*. A project is a set of jobs, tasks, or activities that must be completed in order to achieve specific project objectives, typically which are unique and nonrecurring. A project activity is a discernible task or work function where a particular worker or crew of workers completes a specific item of work within a prescribed time frame. Project activities have certain specific characteristics. Their outcome, work operation, or "job" is measurable. Activities have distinct start and finish times and must consume resources. The responsibility for a project activity must be assignable to a specific manager or work crew. Milestones are key events in the life of a project that measure project progress or phases of project completion such as the finish of an important project activity. Unlike a project activity, a milestone does not have a distinct start and finish time, nor does it consume project resources.

A contractor commences a project with a facility design (i.e., technical drawings and specifications), which provides a very detailed "picture" of the end product or completed facility. The facility is represented as finished or complete, and few details are provided the contractor about how he or she must perform the construction activities to complete the work. The contractor must use the technical descriptions of the completed facility to develop a network plan that executes all aspects of the work to physically construct the facility. The plan must contain all required resources (i.e., time, materials, equipment, and labor) and must be such that the facility is completed on time and within the estimated budget. In the network plan, the contractor must review facility plans and set objectives for the work, analyze the design for alternative construction methods, and develop the network model. The network model includes the steps of defining project activities, ordering the activities in the work sequence, diagramming the sequence in a network format, determining activity durations, calculating the schedule, and adjusting to meet job constraints.

The primary precedence diagramming techniques are the bar chart, activity-on-arrow diagram, activity-on-node diagram, and time-scaled network. The bar chart visually represents activity starts and finishes as well as project progress. Bar charts are easy to generate and interpret, but do not show activity dependency relationships. In activity-on-arrow (AOA) diagrams, activities are represented by arrows, and the start and finish of those activities are represented by nodes. With AOA diagrams, activity start and finish times are clearly depicted as separate events, making it a good tool for showing activity sequence. However, the AOA's use of dummy activities is confusing, and time is not shown graphically. The activity-on-node (AON) diagram represents activities as nodes and the sequence of activities as arrows. The AON eliminates the need for dummy activities (as used in the AOA diagram) and is easy to learn. However, time and project progress are not shown graphically. The time-scaled network shows graphically the activity starts and finishes as well as the order of activities, but is awkward for large, complex projects.

The critical path method, or CPM, is a method developed in the 1950s that uses the network modeling process to schedule projects. The critical path represents the sequence or path of activities that takes the longest to complete. The length of the critical path or sum of all critical activity durations is equal to the minimum project duration. The CPM can be applied to those projects that can be divided into distinct activities with fixed durations and a well-defined sequence of work.

CHAPTER 3   QUESTIONS/PROBLEMS

1. Compare and contrast a project *activity* and a project *milestone*.
2. What are the characteristics of a project activity? In other words, what criteria must be met to identify a project activity?
3. In a construction project, what is the difference between a *procurement* activity and a *management* activity?
4. When a contractor begins to analyze facility design documents (i.e., drawings and specifications) for alternative construction methods, what problem does he or she face?
5. Describe the basic steps of the critical path method where the project planner develops a network plan.
6. Choose three categories of project activities in Section 3.2.3(a) and write a description that explains a "real" example of each category from the construction industry.
7. Assume a 1-yd^3 (0.76-m^3) backhoe can excavate 720 yd^3 (550 m^3) of ordinary soil per 8-hour day where the day actually has 7 hours of productive time with lunch and breaks. If the backhoe is excavating a house basement that measures 30 ft by 40 ft and is 10 ft deep:

   a. How many productive hours are required to excavate the basement assuming average working conditions and 8 hours of productive time per day?
   b. Assuming productivity drops to 80 percent of normal due to difficult soil working conditions, what is the expected duration of the excavation activity?

REFERENCES

See References on page 199.

SCHEDULING LOGIC AND COMPUTATIONS

================================= O B J E C T I V E S =================================

This chapter provides knowledge in the areas of:

❏ Network logic
❏ Logic tables
❏ Activity-on-arrow logic diagrams
❏ Activity-on-node logic diagrams
❏ AON logic calculations—CPM analysis

4.0   OVERVIEW

This chapter describes the construction project in network format. The network model is a graphical diagram of the order of events or sequence in a construction project. The network model is generated following a set of assumptions and/or conventions for diagramming and sequencing. Activity sequences are represented by immediate predecessor lists or logic tables. A logic table provides the activity sequence information including both predecessors and successors for project activities. The network model graphically represents the project in either an activity-on-arrow (AOA) or activity-on-node (AON) format, the most common being the AON format in recent years. The AON network format is used to perform analytical computations by the forward and backward pass method. These computations determine the project's critical path, the project duration, and the start and finish times of all project activities.

4.1   NETWORK LOGIC

In Chapter 3, we learned of the *network model*, which is the sequence of defined activities or the order of events in a construction project represented graphically by one or more of several diagrams including the bar chart, AOA and AON diagrams, and time-scaled network. Calculations are performed on the model to determine the critical path (path of activities that takes the longest to complete in the project), the project duration, the start and finish times of all activities, and the flexible time of noncritical activities. Whereas the emphasis in Chapter 3 was on the development of the network model and a brief description of the process, this chapter provides a detailed explanation of network logic and scheduling computations.

The network model represents project activities, timing, and sequence with some combination of bars, nodes, and/or arrows. Each diagramming technique mentioned above can be modified to show activities, timing, and sequence. However, generally bar charts show activity timing (start and finish times), AOA and AON sequence diagrams show the order in which project activities occur, and time-scaled networks show both timing and sequence (refer to Figures 3.5, 3.6, and 3.8, respectively). The use of a certain type of project scheduling diagram depends on what information needs to be communicated about the project to the various participants, the personal preferences of the project planner, and the current trend of computer software for project scheduling. The last reason is an important one because of the heavy use of computer software for project scheduling, particularly in the construction industry. As stated in Chapter 1, a recent survey of over 40 project management professionals in the construction industry indicated that nearly 100 percent used project management software during the previous 12-month period (Liberatore, Pollack-Johnson, and Smith, 2001). The current trend in project scheduling software is to model projects as bar charts and/or AON diagrams. For these reasons, we will concentrate on these two types of diagrams, although the AOA diagram will be explained as well to provide historical perspective.

Once project activities and durations are defined, often the project planner will use scheduling software to diagram the project rather than sketching or drawing full diagrams by hand. Project scheduling software such as Primavera's Project Planner and Sure-Trak (Primavera Systems, Inc., Bala Cynwyd, Pennsylvania) offers the user the ability to develop the schedule as either a bar chart or an AON diagram. Most project scheduling software products refer to their AON diagrams as "PERT" charts, but this is a misnomer. PERT, or the program evaluation and review technique, is a probability modeling method for scheduling projects that have highly variable activity durations. PERT is applied using AON diagramming, and this probably contributes to the misunderstanding. However, most scheduling software packages do not have PERT capabilities in regard to variable activity durations, and so a more appropriate name to use is AON, sequence, or precedence diagram rather than PERT diagram. More details of the PERT method are provided in Chapter 8.

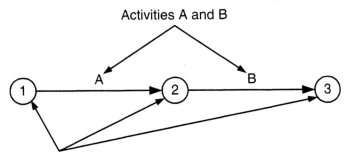

Activities A and B

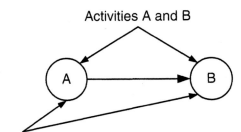

Nodes - In **activity-on-arrow (AOA)** diagrams,
nodes mark the start and finish activities.
Arrows show activities and activity sequence.

Activities A and B

Nodes - In **activity-on-node (AON)** diagrams,
nodes designate the activities. Arrows show
activity sequence only.

FIGURE 4.1 Comparison of AOA and AON diagramming techniques

Because the purpose of this chapter is to examine *scheduling logic and compu-tations* in which activity timing and sequence are important factors, it is necessary to utilize a diagramming technique to best represent those factors. Therefore, we will use the AON diagram to illustrate most points. A sidebar in Chapter 3 explained the use of letters (A, B, C, . . .) and numbers (1, 2, 3, . . .) to designate project activities in scheduling diagrams. This simplistic representation of activities will continue in this chapter because it allows us to concentrate on **project scheduling concepts** rather than individual descriptions of project activities. Once those concepts have been thoroughly developed, we will then look at some examples that are more realistic and represent projects taken from the construction industry.

In the process of developing scheduling diagrams, several assumptions must be made about and conventions followed for the graphical methods commonly used to show network (project) logic. These assumptions apply to all scheduling diagrams discussed here.

1. The sequence of activities—the preceding and succeeding nature of the construction work—is logical. Nodes and arrows are used in diagrams to represent this network logic for both AOA and AON diagrams (see Figure 4.1).

2. Other than the start and finish activities, each activity has a predecessor and successor. Each path of activities from the start of the project flows uninterrupted to the end of the project.

3. For any project activity preceding another activity (e.g., activity A precedes activity B, or A → B), all work for the first or preceding activity must be finished prior to the start of work for the second activity. This is called a *finish-to-start* (FS) activity relationship. You will learn of other more complex relationships, often referred to as the precedence relationships, in Chapter 5. In this chapter, we will assume the FS relationship will be used exclusively for all activity relationships.

4. Network logic diagrams show progress within the project over time going from left to right on the diagram, following the same convention of reading text from left to right as you are doing here. This convention can be clearly seen in the project diagrams in this chapter and in Chapter 3. The left-to-right convention should be observed, if possible, when nodes are placed in a diagram. In other words, numbers and/or letters should generally increase from left to right in the diagrams.

5. In AOA diagrams, only one activity can be defined for any single set (group of two) of nodes. This assumption is illustrated in Figure 4.2, where activity A precedes activity C and activity B also precedes activity C. In Figure 4.2(a), activity A is shown on the arrow and is defined by a start node 1 and a finish node 2. Because activity B also precedes activity C, we can place the beginning of B on start node 1 also but cannot place its end on finish node 2 [Figure 4.2(b)]. AOA diagrams use *dummy activities* to designate sequence in these unique cases, as shown in Figure 4.2(c). This is explained in greater detail in Section 4.3 on AOA logic diagrams.

6. In AON diagrams, arrows are used to represent activity sequence exclusively, and only one arrow can exist between any two activities (Figure 4.1).

7. In AOA and AON sequence diagrams, the start and finish of a project are each represented by single nodes. Depending on the work to be done, a project may start or end with two or more activities. However, the convention here is to place a single beginning activity or node and a single ending activity or node in the diagram. This assures all paths originate from a single node and terminate at a single node whether that node is the end of one or more activities (AOA diagram) or is an activity (AON diagram). In the AON diagram, this convention sometimes requires the use of artificial START and/or FINISH activities (Figure 4.3). These "activities" could be called *milestones* because they have zero duration and consume no resources.

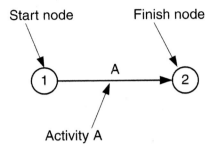

(a) Activity A — Arrow and Nodes

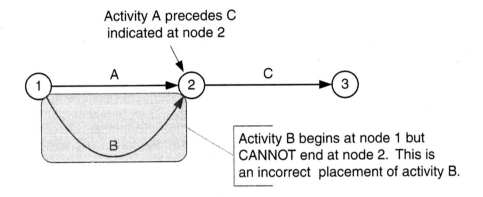

Activity A precedes C
indicated at node 2

Activity B begins at node 1 but
CANNOT end at node 2. This is
an incorrect placement of activity B.

(b) Incorrect Placement of Activity B

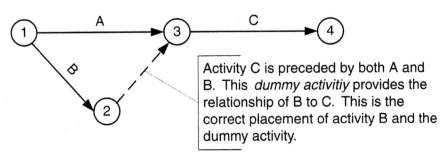

Activity C is preceded by both A and
B. This *dummy activitiy* provides the
relationship of B to C. This is the
correct placement of activity B and the
dummy activity.

(c) Correct Placement of Activity B

FIGURE 4.2   Illustration of AOA activity precedence

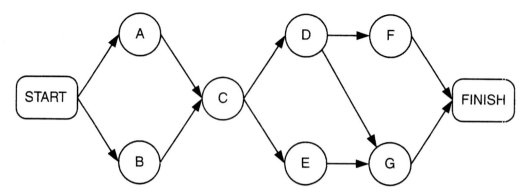

FIGURE 4.3   AON diagram with START and FINISH activities

4.2   LOGIC TABLES

The physical work of a construction project must proceed sequentially, with certain activities preceding others to erect/fabricate/construct a facility. For planning and scheduling purposes, the sequence of project activities is graphically illustrated in sequence diagrams (AOA and AON) where the order of activities is represented by connected arrows and nodes. Before actually drawing the sequence diagrams, the activity sequence must be determined. This sequence is often designated by identifying immediate predecessors or successors for project activities. Immediate predecessor/successor lists for sample projects are given in Tables 3.1 and 3.2 in Chapter 3. The immediate predecessor list and associated logic comments for the AON diagram in Figure 4.3 is provided in Table 4.1.

While immediate predecessor lists provide the necessary information to construct sequence diagrams, often the *logic table* or *precedence grid* is utilized to represent activity sequence. A logic table is a condensed method to display

TABLE 4.1   Immediate Predecessor List for Figure 4.3

ACTIVITY	IMMEDIATE PREDECESSOR	LOGIC COMMENTS
START	—	Precedes activities A and B
A	START	Precedes C
B	START	Precedes C
C	A, B	Precedes D and E
D	C	Precedes F and G
E	C	Precedes G
F	D	FINISH
G	D, E	FINISH
FINISH	F, G	—

TABLE 4.2   Logic Table for Figure 4.3

		A	B	C	D	E	F	G	FINISH
		↓ SUCCESSORS ↓							
	START	X	X						
↓ Predecessors →	A	O		X					
	B		O	X					
	C			O	X	X			
	D				O		X	X	
	E					O		X	
→	F						O		X
	G							O	X

activity sequence information including both predecessors and successors for project activities. The logic table for the AON diagram in Figure 4.3 is shown in Table 4.2.

The shaded boxes in the logic table follow the lead diagonal on the table indicating table cells where activity precedence can't be shown (i.e., A-A, B-B, etc.). In other words, activity A cannot precede or succeed itself and so on. Some textbooks suggest the use of large circles or letters "O" instead of shading the cells. One or both of these methods suffice as long as the lead diagonal is well distinguished and is not used to indicate activity sequence. The activities listed vertically in the logic table are the predecessors of the activities listed horizontally. Thus, the activities in columns are the successors of the activities in rows as identified in Table 4.2. For example, the START activity precedes activities A and B, and conversely activities A and B succeed the START activity. Further, activities D and E precede activity G, and activity G succeeds activities D and E. Notice the FINISH activity does not occur in the predecessor rows since it has no successors, and the START activity does not occur in the successor columns since it has no predecessors.

Logic tables aid in identifying redundant and unnecessary sequence relationships among project activities. An activity relationship is redundant when a relationship path already exists between the anticipated predecessor and successor activity. For example, activity B precedes activity C and C precedes D in Table 4.2. While B must be finished before starting C and thus before starting D, it would be redundant to place a direct relationship between predecessor B and successor D. It is important to remember that logic tables identify immediate predecessors and successors only.

By observing the left-to-right convention for lettering and numbering nodes in activity sequences and thus diagrams, most of the "X" cells in the logic table will occur above the lead diagonal. If the X is marked below the lead diagonal, a counterintuitive logic may exist in the model. This takes place when a higher-numbered/lettered activity precedes one that is a lower letter/number (e.g., activity B preceded by activity F). The project planner may introduce counterintuitive logic when defining project activities and/or assigning activity codes, particularly as project size and complexity increase. It is important to understand that by observing the left-to-right convention for lettering/numbering projects, the resulting logic table should have most, if not all, X marks above the lead diagonal. Any X below the lead diagonal should be carefully examined for accuracy. The logic table provides a clear, unambiguous depiction of activity sequence in preparation for diagramming the logic of the project.

4.3   ACTIVITY-ON-ARROW LOGIC DIAGRAMS

The *activity-on-arrow, or AOA, diagram* was the first project scheduling diagram to be used to apply the CPM and PERT processes. When CPM and PERT were first developed in the late 1950s in the United States and for several decades afterward, the arrow diagram was the method of choice for government agencies awarding large construction contracts and thus for project planners managing these projects. In the development of the AOA diagram (sometimes referred to as the *arrow diagram method, or ADM*), arrows represent activities and nodes represent the start and finish of those activities. Thus, two nodes represent each activity in the AOA method, as shown in Figure 4.2(a). This two-node representation of activities in AOA networks is often called the *i-j node method*. The *i* node, at the tail of the arrow, represents the start of the activity and the *j* node, at the point of the arrow, the finish of the activity. The *i-j* node method was adopted by computer programmers of the time providing the basis for most of the first project scheduling programs. However, sophisticated programs of today do not depend on *i-j* node nomenclature, nor do they use AOA diagrams for network modeling. The AON diagramming technique is more common because of higher flexibility and complications that arise when applying the AOA technique. The basic AOA technique is explained here as well as reasons for the AOA complications as compared with the AON technique. After this explanation of basic AOA, we will shift our emphasis for the remainder of this book to the AON diagramming technique. The AON method is the most common diagramming technique used today and is used almost exclusively by project scheduling computer software for sequence diagramming.

The activity-on-arrow diagramming technique is best illustrated with a relatively simple example of the construction of a foundation for our three-unit townhouse. Basically, the process begins with the foundation area being cleared. Then, while one work crew cuts and bends reinforcing steel for the footer, the foundation corners and spans are accurately located using survey methods. After locating the foundation, the footer can be excavated by a backhoe. Once the excavation is complete and the steel is prepared, the steel can be placed in the footer and concrete can be poured to form the footer. The sequence of this process is given in the logic table in Table 4.3(a) and the predecessor list in Table 4.3(b) and is shown in the AOA diagram in Figure 4.4.

TABLE 4.3(a)   Logic Table for Townhouse Foundation

	ACTIVITIES	↓ SUCCESSORS ↓				
		A	**B**	**C**	**D**	**E**
Predecessors →	A – Clear site		X	X		
	B – Survey site				X	
	C – Cut and bend reinforcing steel					X
	D – Excavate footing					X
	E – Place steel and pour concrete					

TABLE 4.3(b)   Immediate Predecessor List for Townhouse Foundation

ACTIVITY	IMMEDIATE PREDECESSOR
A – Clear site	—
B – Survey site	A
C – Cut and bend reinforcing steel	A
D – Excavate footing	B
E – Place steel and pour concrete	C, D

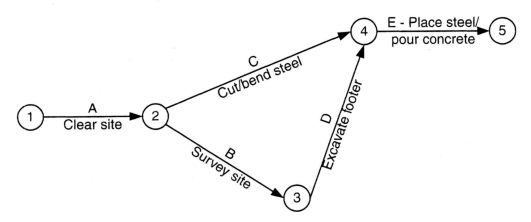

FIGURE 4.4   Townhouse foundation AOA diagram

Suppose that the project planner wanted to add another activity to the project of building the townhouse foundation for more effective communication. The activity—*deliver reinforcing steel*—will occur concurrently with clearing the site (activity A). The logic or sequence of the new activity, called A′, is shown in Table 4.4(a). Notice activity A′ has no predecessor activities and thus can start on the same node 1 as activity A. A complication arises when assigning the finish node for activity A′. From Table 4.4(b), it can be seen that activity A precedes activities B and C, but activity A′ precedes C only. While activities A and A′ can both begin at node 1, they both cannot finish at node 2 for two reasons. First, this

TABLE 4.4(a)   Logic Table with New Activity A′

↓ Predecessors Successors →	A	A′	B	C	D	E
A – Clear site			X	X		
A′ – Deliver reinforcing steel				X		
B – Survey site					X	
C – Cut and bend reinforcing steel						X
D – Excavate footer						X
E – Place steel and pour concrete						

TABLE 4.4(b)   Immediate Predecessor List with New Activity A′

Activity	Immediate Predecessor
A – Clear site	—
A′ – Deliver reinforcing steel	—
B – Survey site	A
C – Cut and bend reinforcing steel	A, **A′**
D – Excavate footing	B
E – Place steel and pour concrete	C, D

would go against the convention we established earlier that only one activity can be defined for any single set of two nodes. Second, node 2 not only defines the finish of activity A; it also defines the start of activities B and C.

Activity A′—*deliver reinforcing steel*—precedes activity C only, not B also. Thus, activities A and A′ are both immediate predecessors of activity C, but A alone is the immediate predecessor of activity B. In arrow diagramming, an activity and its immediate predecessor activity share a single node. The finish or terminal node of the predecessor activity is also the start or initial node of the successor activity. For example, it can be seen in Figure 4.4 that node 2 is both the terminal node for activity A and the initial node for activities B and C. When activity A′ is introduced into the project, it is assigned the same initial node 1 as activity A. The problem is in assigning the terminal node of activity A′. One possibility is to assign node 2 as its terminal node, but by doing this we create the problem encountered in Figure 4.2(b). Also, this action would designate A′ as an immediate predecessor of activities B and C rather than just activity C, which is needed. Another possibility is to draw activity C twice, as shown in Figure 4.5. This complicates the drawing and makes it difficult to track changes and activity/project progress.

The solution to this problem is the use of *dummy activities*. A dummy activity is not really a construction activity at all—it has zero duration and doesn't consume resources. It is applied in AOA diagrams solely to specify sequential relationships for actual activities. In our previous townhouse foundation example, node 2 has been split into nodes 2 and 3, and subsequent node numbering has been increased (Figure 4.6). Activities A and A′ start at the initial node 1. Activities A and A′

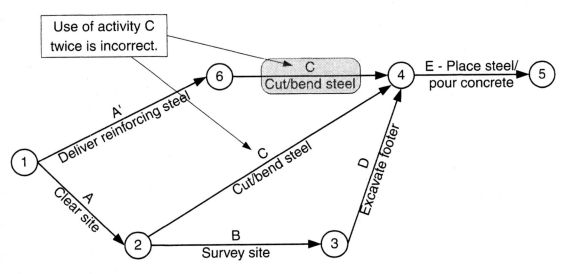

FIGURE 4.5 Incorrect placement of activity C

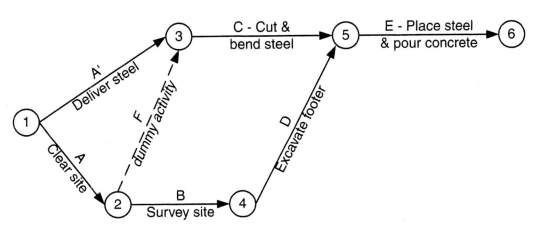

FIGURE 4.6 Townhouse foundation with dummy activity F—AOA diagram

precede C at node 3, and activity A precedes B only at node 2. To show activity A as an immediate predecessor of B also, *dummy activity F* is placed between nodes 2 and 3. The logic for this network in given in Table 4.5.

The use of dummy activities in AOA diagramming is necessary to accurately represent activity sequential relationships. Yet dummy activities can result in a complicated diagram when the number of project activities and the complexity of their relationships increase. This is especially true when there are two or more immediate predecessors for a given activity. Let's look at a simple example to illustrate this point.

TABLE 4.5(a) Logic Table with Dummy Activity F

↓ PREDECESSORS / SUCCESSORS →	A	A'	F	B	C	D	E
A—Clear site	▓		X	X			
A'—Deliver reinforcing steel		▓			X		
F—Dummy activity			▓		X		
B—Survey site				▓		X	
C—Cut and bend reinforcing steel					▓		X
D—Excavate footer						▓	X
E—Place steel and pour concrete							▓

TABLE 4.5(b) Immediate Predecessor List with Dummy Activity F

ACTIVITY	IMMEDIATE PREDECESSOR
A—Clear site	—
A'—Deliver reinforcing steel	—
B—Survey site	A
C—Cut and bend reinforcing steel	A', F
D—Excavate footing	B
E—Place steel and pour concrete	C, D
F—Dummy activity	A

Figure 4.7 represents a small project with seven activities—A through G—and the sequence for the project without the dummy activities is given in the logic table, Table 4.6. Adding the dummy activities necessary to sequence this small project increases the size of the logic table considerably, as shown in Table 4.7. Summarizing the project, activity F is preceded by activities C and D; activity G is preceded by activities A and E; E is preceded by B and C; D is preceded by A and B; and activities A, B, and C have no predecessors. The complexity of the necessary dummy activities—H1, H2, H3, and H4—can easily be seen in the figure and increases the seven-activity project to one with eleven activities. While this simple project was devised expressly to illustrate the complications with AOA diagrams, it exposes a major criticism of the AOA diagramming technique. Additional information on the correct application of dummy activities in AOA diagrams can be found in Wiest and Levy (1977, pp. 20–22).

The use of AOA versus AON diagramming has been and continues to be debated in both industry and academia. On the positive side, AOA diagramming was the first method used and is a long-established method effectively used by experienced planners. Further, AOA provides easy interpretation of activity start and finish times as separate events, and AOA can more easily be

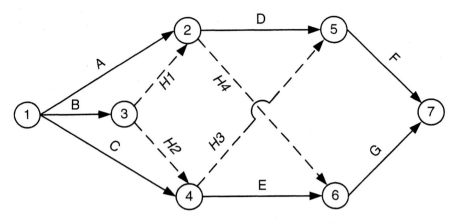

Figure 4.7 Example AOA diagram with multiple dummy activities

Table 4.6 Logic Table without Dummy Activities

		↓ Successors ↓						
		A	B	C	D	E	F	G
→ Predecessors →	A				X			X
	B				X	X		
	C					X	X	
	D						X	
	E							X
	F							
	G							

Table 4.7 Logic Table with Dummy Activities

		↓ Successors ↓									
		A	B	C	H1	H2	H3	H4	D	E	F	G
→ Predecessors →	A							X	X			
	B				X	X						
	C						X			X		
	H1								X			
	H2									X		
	H3										X	
	H4											X
	D									X		
	E											X
	F											
	G											

converted to a time-scaled network using arrow lengths proportional to a time scale. However, as we have demonstrated, the use of dummy activities can be cumbersome and confusing, especially for large projects with complex activity relationships. One of the most compelling reasons AON is more widely applied than AOA in the construction industry is that most contemporary computer software for project scheduling bases its sequence diagramming on the AON method.

4.4   ACTIVITY-ON-NODE LOGIC DIAGRAMS

The node format used in an *activity-on-node,* or *AON, diagram* was first proposed by John W. Fondahl in the United States and Bernard Roy in France. Fondahl initially reported the method in *A Non-Computer Approach to the Critical Path Method for the Construction Industry* (1962). He highlighted the advantages of node format in comparison with the arrow format. However, from the late 1950s to the early 1980s, the arrow format of the AOA diagrams was used primarily for PERT and CPM. During this time, computer programs for project scheduling required arrow logic. The development of personal microcomputers and commercial scheduling software in the early 1980s prompted the growth of the node format. As computers and software progressed, the popularity of the node format for diagramming overtook that of the arrow format. Today, practically all project scheduling software uses the node format.

The AON diagramming method does not utilize dummy activities, the primary complication of the AOA diagram. These are no longer required because the AON method represents activities as nodes and activity sequence as arrows, as was illustrated in Figure 4.3. AON nodes are typically shown as circles (for simple projects, Figure 4.3) and rectangular boxes (for more complex projects—we'll see these later in this chapter). Arrows show only activity sequence, with the beginning of the arrow designating the predecessor activity and the arrow's point designating the successor activity. Let's refer to the example of the three-unit townhouse foundation provided in the previous section on AOA logic diagrams to explain the application of the AON method. The AON diagram for Table 4.3 is shown in Figure 4.8. This figure corresponds to the AOA diagram in Figure 4.4. It is important to note that the sequence diagram in Figure 4.8 follows convention 7 (Section 4.1), which requires the diagram to contain a single beginning node and a single ending node. Activity A is the initial or start activity, and activity E is the final or terminal activity; thus no artificial START or FINISH nodes were necessary. If multiple activities are concurrently taking place at either the start or the finish of a project, it is necessary to use the START or FINISH milestones.

Adding activity A'—*Deliver reinforcing steel*—to occur concurrently with activity A—*Clear site*—is a simple process using the AON format (Figure 4.9). Because the project now begins with the concurrent activities A and A', a START milestone is required. In Figure 4.9, it is easy to see the sequence of activities, particularly activity A preceding activities B and C while the newly inserted activity

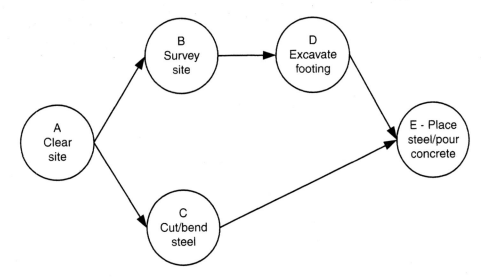

Figure 4.8 Townhouse foundation AON diagram

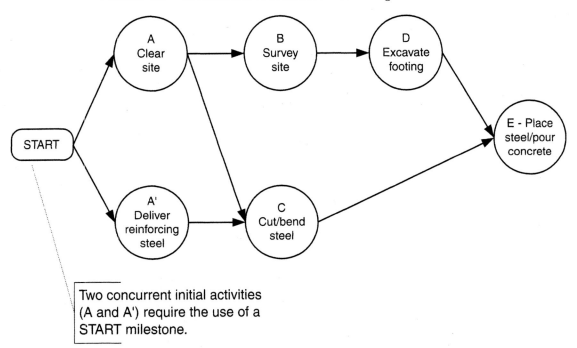

Two concurrent initial activities
(A and A') require the use of a
START milestone.

Figure 4.9 AON diagram for Table 4.4

A' precedes activity C only. Figure 4.10 illustrates the use of the AON format for the project in Table 4.6. Both START and FINISH milestones are used in Figure 4.10. When the AON diagram in Figure 4.10 is compared with its counterpart AOA diagram in Figure 4.7, the advantage of the AON method becomes clear.

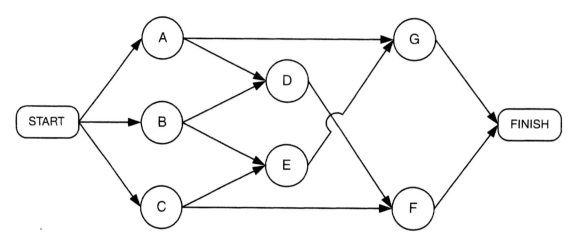

FIGURE 4.10   AON diagram for Table 4.6

4.5   AON Logic Calculations—CPM Analysis

4.5.1   Time-Constrained versus Resource-Constrained Scheduling

In Chapter 3, you learned that project *activities* consume resources (i.e., time, material, equipment, labor, money). The careful analysis of resources is a key component of good project management. The project manager must determine the sequence of activities that are most critical for the successful completion of the project and allocate the appropriate resources for this sequence. Of great importance is the ability to determine and make adjustments for resource conflicts, especially for those activities that are most critical. Resource conflicts result from resource constraints. A resource is constrained when the limited availability of that resource (e.g., equipment, material, labor) occurs during a time in the schedule where the resource is required by two or more concurrent project activities.

In order to determine resource conflicts during a project, the project manager must first determine the project schedule constrained by time only—*time-constrained scheduling*. In other words, no resource constraints other than those applied by activity durations and their relationships are considered. Once the schedule has been determined with start and finish times of each project activity, the project manager can then begin to analyze the schedule as constrained by resources other than time—*resource-constrained scheduling*—and make adjustments to resolve the conflict. In this chapter, we concentrate on time-constrained scheduling, and in Chapter 6, we focus on resource-constrained scheduling.

4.5.2 CALENDAR AND ORDINAL DATES IN SCHEDULING

Calendar dating, or designating the days the project will occur as dates on a yearly calendar, is the commonly used method in the application of project scheduling to construction projects. The determination of the project schedule based on calendar dates provides all concerned parties with the actual days and dates of the planned project activities. Important dates can be identified during the project such as delivery deadlines and completion milestones. *Ordinal dating,* or ordering the project days in numerical order, is used to simplify the process of designating project days. In its simplest form, ordinal dates number the first day of the project as day 1, the second day as day 2, and so on. The ordinal dating format may be used to communicate certain project objectives by concentrating on project activities rather than on the actual weekday or calendar date when those project activities transpire. However, this is not a common practice in industry—calendar dating is the norm.

Ordinal dating is particularly useful for the purpose of understanding the computational technique used in the critical path method. An understanding of these basic scheduling computations is essential for properly interpreting computer outputs and for making the necessary adjustments in project schedules. While this technique—the *forward and backward pass method*—is mathematically straightforward, the learner must be able to closely follow the day-to-day or week-to-week (depending on the time unit used) behavior of the schedule. The process is more easily understood while not encumbered by calendar dating, in which a normal workweek is typically assumed and weekends are not workdays. Because a primary intention of this text is to present the concepts of project planning and scheduling and these concepts can be more readily understood with ordinal dating, this dating format is used predominately throughout the book. Conversion of ordinal to calendar dates requires the actual project start date, the number of normal workdays per week (typically 4, 5, 6, or 7), and the holidays to be included. It should be noted here that most project scheduling computer software allows the use of and conversion between ordinal and calendar dating.

4.5.3   FORWARD AND BACKWARD PASS COMPUTATIONAL METHOD

The critical path method involves defining project activities, putting them in sequence, diagramming the network, determining activity durations, and calculating the project schedule. Calculating the schedule involves setting a beginning date for the project and numerically determining the start and finish times of individual project activities, relative to their sequence and durations. The algorithm, or analytical steps used to calculate the schedule, is the **forward and backward pass method.** This method is explained assuming an AON project network format rather than an AOA format for purposes noted earlier (see Sections 4.3 and 4.4). The assumptions and conventions presented in Sections 4.1 of this chapter are followed for the forward and backward pass method.

The forward and backward pass method is quite definitive as a title because all scheduling computations involve first a forward and then a backward pass. The process begins by specifying a project start or *initial time*, typically at time equals zero (i.e. the first activity begins at the start of day #1). The *forward pass* proceeds sequentially from the beginning to the end of the project along each network path to include all project activities. Computations on the forward pass yield the expected earliest time that each activity can start and finish and the expected project duration (finish of the last or latest project activity). Let's remember that the project scheduling process is one where a model is developed to represent the project and the results of the model computations are "expected" to occur based on the model. The actual time of activities as they occur during a project may vary from the expected or modeled time depending on the accuracy of our estimates of activity duration and sequence.

Using the finish time of the last project activity or other longer *terminal time* as the project duration (i.e., specific future target date), the *backward pass* proceeds sequentially from the end to the beginning of the project along all network paths to include all activities. Backward pass computations yield the expected latest allowable start and finish times for each activity. From the early and late start and finish times for each activity, the float time of each activity is computed and the critical and noncritical paths through the network are identified.

Both the forward pass and the backward pass are two-step procedures, which are applied to individual project activities, in topological sequence. In the following formulas, activities are indexed as predecessor activity i with successor activity j. A topological sequence is one where predecessor and successor activities, i and j, are arranged in ascending order. The following nomenclature is used for formulas and for discussion of forward and backward pass computations:

D_j = duration time for activity j
ES_j = early start time for activity j
EF_j = early finish time for activity j
LS_j = late start time for activity j

LF_j = late finish time for activity j

FS_{ij} = finish-to-start constraint

The finish-to-start constraint, FS_{ij}, allows the project planner to place a lag time between the finish of a predecessor activity and the start of its successor activity. In this chapter, we will assume that this constraint or lag time is equal to zero following the conventions given in Section 4.1 of this chapter. In Chapter 5, the FS_{ij} and other advanced activity relationship constraints are fully detailed and their uses illustrated.

Forward pass

Step 1. Compute the early start time, ES_j, of the activity (j) in question.

$$ES_j = \underset{\text{all } i}{\text{maximum}} \begin{bmatrix} \text{initial time, } 0 \\ EF_i + FS_{ij} \end{bmatrix} \tag{4.1}$$

Step 2. Compute the early finish time, EF_j.

$$EF_j = ES_j + D_j \tag{4.2}$$

Backward pass

Step 1. Compute the late finish time, LF_i, of the activity (i) in question.

$$LF_j = \underset{\text{all } j}{\text{minimum}} \begin{bmatrix} \text{terminal time} \\ LS_j - FS_{ij} \end{bmatrix} \tag{4.3}$$

Step 2. Compute the late start time, LS_i.

$$LS_i = LF_i - D_i \tag{4.4}$$

The initial time is the project start time, normally set at zero, and the terminal time is the project finish time, normally equal to the early finish time of the last activity in the topological sequence. Once the early and late times have been determined via the forward and backward pass method, the activity float time (Equation 4.5) and project duration (Equation 4.6) can be determined.

$$\text{Activity float} = \begin{bmatrix} LF_j - EF_j \\ \text{or} \\ LS_j - ES_j \end{bmatrix} \tag{4.5}$$

$$\text{Project duration, } T_s = \begin{bmatrix} \text{terminal time} \\ \text{or} \\ EF_{\text{last activity}} = LF_{\text{last activity}} \end{bmatrix} \tag{4.6}$$

ACTIVITY FLOAT (SLACK): There is some confusion in project management literature regarding the meanings of the terms *activity float* and *activity slack*. Rather than attempt to compare and clarify usage of terms broadly, we will concentrate on the three most common terms used to define **activity float.**

Path or Total Float—Path or total float is defined as the amount of time by which an activity can be delayed if all preceding activities take place at their earliest possible start times and the following activities are allowed to wait until their latest possible start times.

Activity or Free Float—Activity or free float is defined as the amount of float time available if all preceding activities take place at their earliest possible start times and the following activities also take place at their earliest possible start times.

Independent Float—Independent float is defined as the amount of float time available if the preceding event takes place at its latest possible start time and the following activities take place at their earliest possible start times.

CRITICAL PATH: The *critical path* is the network path with the least path or total float. In the case where the project duration is equal to the early or late finish of the last or latest activity, the float along the critical path is equal to zero—this is called a *zero-float* convention.

Forward and Backward Pass Example: Let's illustrate the forward and backward pass computational method with an example project. This small project has eight activities identified by letters (A, B, C,...), as shown in the AON network diagram in Figure 4.11. The project activities, activity durations, and immediate predecessors are given in Table 4.8. The forward and backward pass computations yield early and late start and finish times for each activity. All these data must be displayed for each activity on the diagram to properly track and perform the computations manually. A legend is often used to clarify data placement on the diagram. Three main types of legends, as shown in Figure 4.12, are common in order to represent project activities in a diagram.

The forward pass calculations are explained step by step as follows, based on the assumption that the project starts at an initial time = zero.

TABLE 4.8 Forward and Backward Pass Example Project

ACTIVITIES	DURATION, WEEKS	IMMEDIATE PREDECESSORS
A	3	—
B	5	A
C	4	A
D	2	B, C
E	6	B, C
F	4	D
G	2	E
H	1	F, G

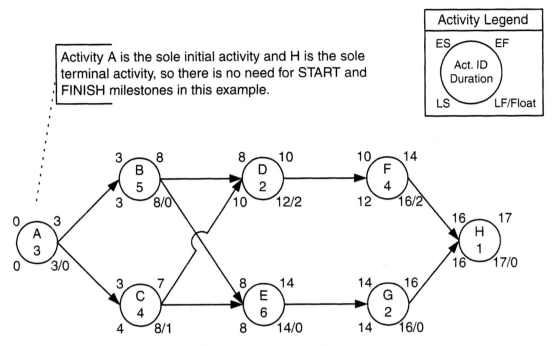

Activity A is the sole initial activity and H is the sole terminal activity, so there is no need for START and FINISH milestones in this example.

Activity Legend

FIGURE 4.11   Forward and backward pass AON diagram

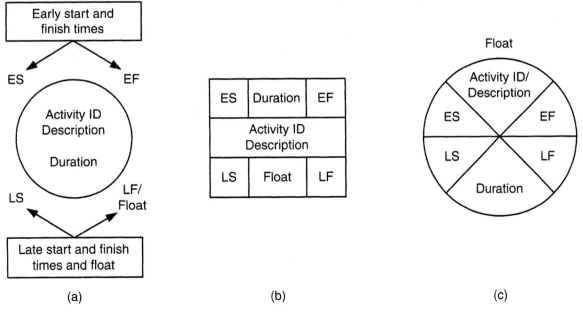

FIGURE 4.12   Project activity diagramming legends

(a)   (b)   (c)

Activity A ES_A = initial time = 0
$EF_A = ES_A + D_A = 0 + 3 = 3$

Activity B $ES_B = EF_A = 3$
$EF_B = ES_B + D_B = 3 + 5 = 8$

Activity C $ES_C = EF_A = 3$
$EF_C = ES_C + D_C = 3 + 4 = 7$

Activity D $ES_D = \max\begin{bmatrix} EF_B \\ EF_C \end{bmatrix} = EF_B = 8$

$EF_D = ES_D + D_D = 8 + 2 = 10$

Activity E $ES_E = \max\begin{bmatrix} EF_B \\ EF_C \end{bmatrix} = EF_B = 8$
$EF_E = ES_E + D_E = 8 + 6 = 14$

Activity F $ES_F = EF_D = 10$
$EF_F = ES_F + D_F = 10 + 4 = 14$

Activity G $ES_G = EF_E = 14$
$EF_G = ES_G + D_G = 14 + 2 = 16$

Activity H $ES_H = \max\begin{bmatrix} EF_F \\ EF_G \end{bmatrix} = EF_G = 16$
$EF_H = ES_H + D_H = 16 + 1 = 17$

Project duration, $T_S = EF_{\text{last activity}} = EF_H = 17$

The early start of activity B above is determined from the early finish time of the preceding activity A using Equation 4.1 where $ES_B = EF_A = 3$ weeks. In Equation 4.1, the magnitude of the early finish of the preceding activity is transferred to the start of the successor. In the case of activities A and B, this magnitude is 3 weeks. It is important to understand that the transfer of the value of 3 in this case actually indicates a point in time rather than a time duration. In other words, the early finish of activity A and the early start of activity B are at the end of week 3, which can also be described as the beginning of week 4. However, it is best to realize that the value determined as the early start time for any given activity is actually at the end of the time duration for that particular time unit (i.e., the end of the third week for ES_B). While it may seem logical to assign a value of 4 weeks for ES_B due to it actually starting at the beginning of week 4, this would create an obvious error in the computations for the forward and backward pass method.

The backward pass calculations are explained step by step as follows, based on the assumption that the project terminal time is equal to the project duration calculated in the forward pass as the early finish time of the last activity, $EF_H = 17$ weeks.

Activity H $LF_H = EF_H = 17$
$LS_H = LF_H - D_H = 17 - 1 = 16$
Activity float $= LF_H - EF_H = 17 - 17 = 0$

Activity G $LF_G = LS_H = 16$
$LS_G = LF_G - D_G = 16 - 2 = 14$
Activity float $= LF_G - EF_G = 16 - 16 = 0$

Activity F $LF_F = LS_H = 16$
$LS_F = LF_F - D_F = 16 - 4 = 12$
Activity float $= LF_F - EF_F = 16 - 14 = 2$

Activity E $LF_E = LS_G = 14$
$LS_E = LF_E - D_E = 14 - 6 = 8$
Activity float $= 0$

Activity D $LF_D = LS_F = 12$
$LS_D = LF_D - D_D = 12 - 2 = 10$
Activity float $= 2$

Activity C $LF_C = \min\begin{bmatrix} LS_D \\ LS_E \end{bmatrix} = LF_E = 8$

$LS_C = LF_C - D_C = 8 - 4 = 4$
Activity float $= 1$

Activity B $LF_B = \min\begin{bmatrix} LS_D \\ LS_E \end{bmatrix} = LF_E = 8$

$LS_B = LF_B - D_B = 8 - 5 = 3$
Activity float $= 0$

Activity A $LF_A = \min\begin{bmatrix} LS_B \\ LS_C \end{bmatrix} = LF_B = 3$

$LS_A = LF_A - D_A = 3 - 0 = 3$
Activity float $= 0$

Let's summarize the data in tabular form in order to draw some conclusions about our project. The data are presented in Table 4.9.

TABLE 4.9   Forward and Backward Pass Results

ACTIVITIES	EARLY START	EARLY FINISH	LATE START	LATE FINISH	FLOAT
A	0	3	0	3	0
B	3	8	3	8	0
C	3	7	4	8	1
D	8	10	10	12	2
E	8	14	8	14	0
F	10	14	12	16	2
G	14	16	14	16	0
H	16	17	16	17	0
		Project duration $= 17$			

With the assumption of the terminal time equal to the project duration (i.e., 17 weeks), a *zero-float convention* is applicable to determine the critical path of the project. In other words, those activities with float time equal to zero are the critical activities—A, B, E, G, H. These activities occur along one specific path in the network; therefore, that activity path is the critical path. Activities C, D, and F are noncritical designated by their available float time, specifically their total or path float time (according to the definitions given previously). The total or path float for activity C is 1 week and is 2 weeks for activities D and F. The activity or free float for activity C is 1 week, F is 2 weeks, and D is 0 weeks. Activity C has 1 week and D and F have 0 weeks independent float.

4.6   SUMMARY

The network model graphically represents the sequence of project activities or order of events in a construction project often by one or more of two types of diagrams— the bar chart and the activity-on-node (AON) diagram. Calculations are performed on the model to determine the critical path (path of activities that takes the longest to complete in the project), the project duration, the start and finish time of all activities, and the float time of noncritical activities. The diagramming process follows several assumptions and conventions related to the proper display of network logic, activity predecessor/successor relationships, project time flow conventions, and proper use of nodes and arrows for diagrams.

Activity sequence or logic is represented in a condensed format with logic tables or precedence grids. Logic tables are used to produce either activity-on-arrow or activity-on-node diagrams, which specifically show activity sequence. In AOA diagrams, arrows represent activities and nodes represent the start and finish of those activities. In AON diagrams, activities are represented as nodes and activity sequence as arrows. The AON diagramming method is the most common of the two in recent years.

Once the AON diagram has been developed, an analytical algorithm called the forward and backward pass method is used to calculate the schedule. The forward and backward pass method begins by specifying a project start time and proceeds sequentially from the beginning to the end of the project along each network path to include all project activities. Computations on the forward pass yield the expected earliest time that each activity can start and finish and the expected project duration (finish of the last or latest project activity). The backward pass proceeds sequentially from the end to the beginning of the project along all network paths to include all activities. Backward pass computations yield the expected latest allowable start and finish times for each activity. From the early and late start and finish times for each activity, the float time of each activity is computed and the critical and noncritical paths through the network are identified.

CHAPTER 4   QUESTIONS/PROBLEMS

1. Group Activity (four to six-member teams) AOA versus AON network diagramming

   - Discuss the advantages and disadvantages of each network diagramming technique.
   - Give reasons for the expanding use of AON diagramming for project scheduling in recent years.

2. Why can only one activity be defined for any single set of nodes (group of two) in AOA diagrams?

3. What is the purpose of assigning a single START (initial) or FINISH (terminal) node in the project diagramming process when the project either begins or is completed by multiple activities?

4. For the logic table below, generate an AOA diagram and an AON diagram using the appropriate assumptions and conventions provided in this chapter.

Logic Table for Problem 4

		↓ SUCCESSORS ↓						
		A	B	C	D	E	F	G
→ Predecessors →	A		X	X				
	B				X	X		
	C					X		
	D						X	
	E						X	
	F							X
	G							

5. For the AON diagram below, generate a logic table in the format used in Problem 4 above.

6. For the following table of project activities, immediate predecessors, and activity durations, generate a logic table, accurately draw an AON precedence diagram using the activity legend indicated below, and perform the forward and backward pass computations. Determine the early start, early finish, late start, late finish, and total float of each activity and project duration.

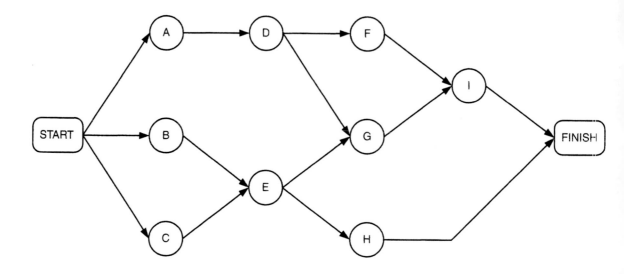

Activity	Duration, days	Immediate Predecessor(s)
A	4	---
B	5	---
C	8	A
D	3	B, C
E	2	B
F	4	D
G	3	F
H	2	E, G

Activity Legend

ES — EF

Act. ID
Duration

LS — LF/Float

REFERENCES

See References on page 199.

PRECEDENCE NETWORKING

━━━━━━━━━━━━ O B J E C T I V E S ━━━━━━━━━━━━

This chapter provides knowledge in the areas of:

- ❏ Activity relationships
- ❏ Precedence networking computations
- ❏ Precedence networking—diagram solution method
- ❏ Planned versus actual project schedule method
- ❏ Project scheduling computer applications

5.0 OVERVIEW

This chapter describes an extension of the critical path method (CPM) from Chapter 4. This extension, called precedence networking (PN), allows the use of complex relationships between activities. In addition to the traditional finish-to-start relationship covered in Chapter 4, this chapter includes descriptions of the start-to-start and finish-to-finish relationships and the associated lag time. The chapter follows with a discussion of project activity splitting. Computations for the PN method are explained and detailed through examples. Finally, the chapter provides an overview of Primavera's SureTrak project management software. Use of the software is explained using an example project.

5.1   ACTIVITY RELATIONSHIPS

In Chapter 3, we defined the project schedule as a *network model* and the project activity sequence or ordered system of events as a *precedence network*. The precedence network, often referred to as the precedence diagramming method (PDM), is a CPM scheduling process whereby project activities are graphically represented as nodes and activity relationships as arrows. The activity-on-node (AON) diagramming method rather than the activity-on-arrow (AOA) method is more commonly used in this process. AON diagramming is continued in this chapter to illustrate an important extension of the CPM process presented in Chapter 4.

The following assumption was made in Chapter 4 regarding activity sequencing for the CPM process. For any project activity preceding another activity (e.g., activity A precedes activity B, or A → B), all work for the first or preceding activity must be finished prior to the start of work for the second activity. This is described as a finish-to-start (FS) activity relationship. In 1964, an extension of the CPM process, specifically related to *advanced activity relationships*, was introduced in an IBM computer users manual (IBM, 1968). The document refers to the extension as *precedence diagramming*, hence the term *precedence diagramming method*. In this method, the FS activity relationship assumption is extended to include several other activity relationships. These advanced relationships provide the project planner with much more flexibility in modeling a construction project.

Precedence networking differs from CPM in that activity relationships in PN can be represented in a more complex form than the basic finish-to-start relationship. Often in large and/or complicated construction projects, more complex relationships exist and must be represented. PN was developed to enact additional types of activity relationships or constraints. The most common PN constraints include the traditional *finish-to-start* relationship used in CPM and two others—the *start-to-start* (SS) and the *finish-to-finish* (FF) relationships. Not only does PN permit these alternative activity relationships, but it also allows for time-lag quantities to be assigned to relationships. An additional type of relationship is the *start-to-finish* (SF) relationship wherein a time lapse is designated between the start of the predecessor activity and the finish of the successor activity. However, this activity relationship has limited use in almost all circumstances in the construction industry and will not be discussed further here.

5.1.1   FINISH-TO-START ACTIVITY RELATIONSHIP

The finish-to-start activity relationship, where one activity must finish before the start of its successor, is traditional for CPM. It is the most common relationship in construction scheduling and is predominately used to represent activity sequence. With this relationship defined between the predecessor activity A and successor activity B, activity B can begin only after activity A is complete. Figure 5.1 provides graphical representation, in both AON and bar chart format, of the finish-to-start

(a) FS relationship — AON diagram

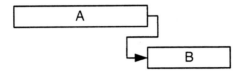

(b) FS relationship — bar chart

FIGURE 5.1   Finish-to-start (FS) activity relationship

relationship. Note the designation of the FS relationship on the arrow of the AON diagram, a common coding style when multiple activity relationships are used in a schedule.

5.1.2   LAG (LEAD) TIME

A lag quantity is a time period that separates or designates a waiting period between two activities. The time lag is associated with the relationship arrow between two activities—a predecessor and successor. In the case of a finish-to-start relationship between two activities, as in Figure 5.1, lag time may be assigned to delay the start of activity B after the completion of activity A. Up to this point, we have assumed that activity B will actually begin immediately after the completion of A with no delay, as in Figure 5.1. Under this assumption, a delay in the start of activity B after the completion of A can occur in one of only two ways: (1) If A and B are critical activities (on the critical path), a delay in the start of activity B extends the critical path time and thus the project duration; or (2) if activities A and B are not on the critical path, B can be delayed by using available float time. Neither of these methods may be desirable options for the project planner.

Assigning lag time between activities A and B allows the start of activity B to be delayed without increasing project duration or using float time. Figure 5.2 illustrates the use of lag time in a simple example. This series of activities involves the pouring of concrete in forms for a retaining wall, curing the concrete, and stripping off the forms. These activities can be shown as in Figure 5.2(a) with

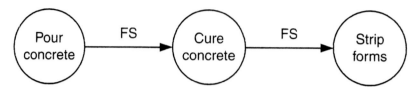

(a) Typical FS relationship sequence

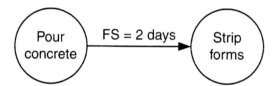

(b) FS relationship with lag time

FIGURE 5.2   Lag time—finish-to-start relationship

a traditional finish-to-start relationship between three activities. This may seem valid, but the activity *Cure concrete* does not require resources other than time and is used here to allow time between the concrete pour and the removal of forms. With a lag time assigned between these two activities, as in Figure 5.2(b), the *Cure concrete* activity is no longer required and yet the scheduling logic remains the same.

Lag time between two activities is designated with the relationship code (i.e., FS for finish to start) set equal to the lag quantity. For example, the lag time to cure concrete in the example in Figure 5.2 is 2 days. Thus, the designation is "FS = 2 days" and is typically shortened to "FS = 2" with consistent time units used in the model. If no lag time is placed on the relationship, a lag equal to zero is implied. It is recommended to include the relationship code and lag time (especially when nonzero) on the relationship arrow when solving the PN schedule manually. This labeling clearly communicates the relationship type and lag time, if any, maintaining consistency and increasing accuracy during the manual solution.

Lag times are typically positive values indicating a delay or waiting period between two activities. However, there may be a need to overlap the two activities. Using the FS relationship as an example, the start of the successor activity could overlap by some predetermined time the finish of the predecessor activity. In this circumstance, the lag time is assigned a negative value and more appropriately called a *lead* or *lead time*. While surely not as common as the assignment of lag time between two activities, lead time provides the project planner flexibility to model the project more realistically.

5.1.3 START-TO-START ACTIVITY RELATIONSHIP

The start-to-start (SS) relationship is applied to schedule the start of a successor activity at the same time as or after some lag time to the start of the predecessor activity (see Figure 5.3). As an example of the start of two activities (predecessor → successor) occurring at the same time, let's examine a concrete pour of a slab on grade. If activity A is *Pour concrete for slab* and activity B is *Level and smooth concrete*, activity B obviously can't begin until A has started but should begin immediately after the start of A. Therefore, the start-to-start relationship between activities A and B is assigned a lag time equal to zero (SS = 0).

A more common application of the start-to-start relationship is with a lag time greater than zero. Let's use an example from our three-unit townhouse (Appendix A) to depict this application. After the walls of the two-story townhouses have been framed, the exterior faced, the electrical/plumbing/HVAC rough-ins completed, and the roof system installed, one of the next steps in the project is to install insulation in the walls of the three units. This insulation activity must be started and at least partially completed prior to starting the drywall installation process. Assuming it takes 2 days to complete the insulation activity in the townhouse complex, installation of drywall can begin after the first day of insulation. In other words, about half the rooms in the townhouse complex will be insulated at the end of 1 day so that the drywall installation can begin in one or more of those rooms at the beginning of the second day. These two predecessor/successor activities are assigned a start-to-start relationship with a lag time of 1 day (SS = 1 day, as seen in Figure 5.3).

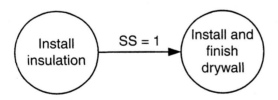

(a) SS relationship — AON diagram

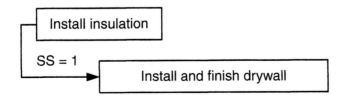

(b) SS relationship — bar chart

FIGURE 5.3 Start-to-start (SS) activity relationship

5.1.4 FINISH-TO-FINISH ACTIVITY RELATIONSHIP

The finish-to-finish (FF) relationship places no restriction on activity start times, but rather requires that the finish time of the successor activity be no earlier than the finish time of the predecessor activity. For example, the final cleanup of a construction area immediately precedes inspection of the constructed facility. But realistically, the inspection does not have to wait until the cleanup has been completed. Inspection and cleanup can occur concurrently (run in parallel) as long as the cleanup of any portion of the construction area is followed by the inspection. It is more accurate to tie the finish times of the cleanup and inspection together with a finish-to-finish relationship equal to zero (FF = 0).

As with the start-to-start activity relationship, the finish-to-finish lag time is commonly greater than zero. For example, the painting of the rooms in the three-unit townhouse must precede carpet installation. Like the cleanup and inspection example above, a finish-to-finish relationship between painting and carpet installation is a realistic representation of the schedule for these two activities. However, it is important that sufficient lag time be allotted between the finish of painting and the finish of carpet installation. This lag time not only allows the paint crew(s) to stay ahead of the carpet installers but also allows for paint drying time. Thus, a 2-day lag time (FF = 2 days) may be appropriate, as shown in Figure 5.4.

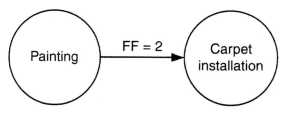

(a) FF relationship — AON diagram

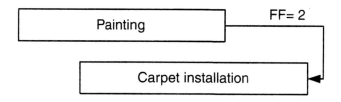

(b) FF relationship — bar chart

FIGURE 5.4   Finish-to-finish (FF) activity relationship

5.1.5 ACTIVITY SPLITTING

Activity splitting is dividing an activity into two or more subactivities with some amount of idle time between the work of the subactivities. Activities are split to increase scheduling flexibility and to enhance the model's representation of the actual project. Most commonly, an activity is split when a specific finish time constraint is greater (at a later time) than the original scheduled finish time placed on the activity. This may be due to equipment or materials availability and/or the need to adjust labor requirements.

Whenever activity constraints result in a later (specified) finish time for the activity, the difference in early start and finish times (or late start and finish times) is greater than the activity duration. If you recall, Equation 4.2 in Chapter 4 states that $EF_j = ES_j + D_j$; therefore $D_j = EF_j - ES_j$. When this occurs, the activity must be split in order to maintain the computational accuracy of the mathematical procedure and preserve scheduling continuity.

The PN computational procedure is considerably more complex when splitting of activities is allowed. This is particularly true when all activities in a project schedule are allowed to be split. Also, it is common for splitting to occur due to an activity finish time being specified at a later time than initially calculated in the schedule. In other words, splitting is a result of the project planner modifying the finish times of one or more activities after the schedule has been determined. Therefore, for the purpose of this text, we will assume that activity splitting is not allowed. For a detailed explanation of the algorithm (set of mathematical steps) for PN scheduling computations with activity splitting allowed, refer to Moder, Phillips, and Davis (1983, pp. 108–111, 126–132).

5.2 PRECEDENCE NETWORKING COMPUTATIONS

The computational method used in precedence networking is more complex than the previous CPM method in which only FS activity relationships were allowed. Advanced activity relationships (i.e., SS and FF) complicate the forward and backward pass. On the forward pass, the SS relationship is of particular concern due to the connection of activity start times rather than the simple transferring of the early finish time of a predecessor activity to the early start time of a successor activity. Similarly, on the backward pass the FF relationship is troublesome. These complications can be seen in the equations for the forward and backward pass below and are illustrated in the examples provided.

As with the CPM computational procedure, the PN procedure uses a sequential forward and backward pass through the network to calculate the early start-to-finish and late start-to-finish times for each activity. From these values, float time, the critical path(s), and the project duration are determined. Also similar to CPM, the forward and backward pass are applied to individual project activities in topological sequence, activities are indexed as predecessor activity i with successor activity j, and the initial (forward pass) and terminal (backward pass) times are specified to begin each pass. The following nomenclature is used for formulas and the discussion of PN computations:

D_j = duration time for activity j
ES_j = early start time for activity j
EF_j = early finish time for activity j
LS_j = late start time for activity j
LF_j = late finish time for activity j
FS_{ij} = finish-to-start constraint between activities i and j
SS_{ij} = start-to-start constraint between activities i and j
FF_{ij} = finish-to-finish constraint between activities i and j

Forward pass

Step 1:   Compute the early start time, ES_j, of the activity (j) in question.

$$ES_j = \max_{\text{all } i} \begin{bmatrix} \text{initial time, 0} \\ EF_i + FS_{ij} \\ ES_i + SS_{ij} \\ EF_i + FF_{ij} - D_j \end{bmatrix} \tag{5.1}$$

Step 2:   Compute the early finish time, EF_j.

$$EF_j = ES_j + D_j \tag{5.2}$$

Backward pass

Step 1:   Compute the late finish time, LF_i of the activity (i) in question.

$$LF_i = \min_{\text{all } j} \begin{bmatrix} \text{Terminal time} \\ LS_j - FS_{ij} \\ LF_j - FF_{ij} \\ LS_j - SS_{ij} + D_j \end{bmatrix} \tag{5.3}$$

Step 2:   Compute the late start time, LS_i.

$$LS_i = LF_i - D_i \tag{5.4}$$

The initial time is the project start time, normally set at zero, and the terminal time is the project finish time, normally equal to the early finish time of the last activity in the topological sequence. Once the early and late times have been determined, utilizing the forward and backward procedures, the activity float times and project duration can be determined, as follows:

$$\text{Activity float} = \begin{bmatrix} LF_j - EF_j \\ \text{or} \\ LS_j - ES_j \end{bmatrix} \quad (5.5)$$

$$\text{Project duration} = \begin{bmatrix} \text{terminal time} \\ \text{or} \\ EF_{\text{last activity}} = LF_{\text{last activity}} \end{bmatrix} \quad (5.6)$$

The critical path is determined, as in the CPM and PERT methods, by the activity sequence with the minimum float. In the case where the project duration is equal to the early or late finish of the last activity, the float along the critical path is equal to zero.

In order to determine early and late start and finish times for each activity for PN, greater care must be observed during the manual solution of the schedule. Project scheduling computer software makes a simple process of these computations, but as in Chapter 4, valuable knowledge and understanding of the scheduling process is gained by performing the computations manually. For this purpose, let's examine the following example.

An AON diagram of a sample project schedule is provided in Figure 5.5 with nine activities and eleven relationship arrows. Seven arrows are traditional finish-to-start activity relationships, and four are either start-to-start or finish-to-finish relationships. Of the seven traditional FS relationships, six have a lag time equal to zero (FS = 0), and the remaining one between activities B and E has a lag time equal to 2 ($FS_{BE} = 2$). Note those relationship arrows with FS = 0 have been omitted from the diagram in Figure 5.5 in order to reduce clutter on the diagram.

FORWARD PASS

The forward pass calculations for the project in Figure 5.4 are explained below. All steps where SS and FF relationships are used as well as where $FS_{BE} > 0$ are denoted at the right of the equations. The project is assumed to start at an initial time equal to zero.

Activity A	$ES_A = \text{initial time} = 0$
	$EF_A = ES_A + D_A = 0 + 2 = 2$
Activity B	$ES_B = EF_A = 2$
	$EF_B = ES_B + D_B = 2 + 5 = 7$
Activity C	$ES_C = EF_A = 2$
	$EF_C = ES_C + D_C = 2 + 6 = 8$

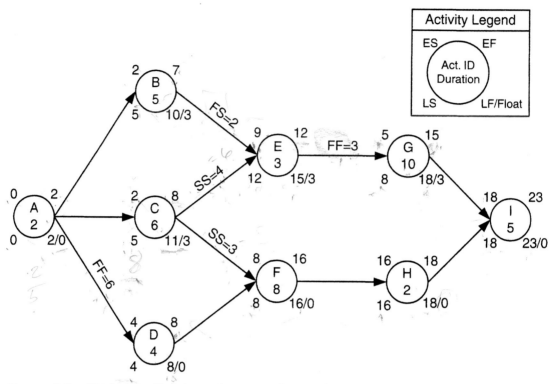

FIGURE 5.5   PN forward/backward pass—advanced activity relationships

Activity D

$$ES_D = EF_A + FF_{AD} - D_D =$$
$$= 2 + 6 - 4 = 4 \qquad \leftarrow FF = 6 \text{ relationship}$$

$$EF_D = ES_D + D_D = 4 + 4 = 8$$

or

$$EF_D = EF_A + FF_{AD} = 2 + 6 = 8$$

Activity E

$$ES_E = \max \begin{bmatrix} EF_B + FS_{BE} \\ ES_C + SS_{CE} \end{bmatrix} =$$

$$= \max \begin{bmatrix} 7 + 2 \\ 2 + 4 \end{bmatrix} = 9 \qquad \leftarrow FS = 2 \text{ and } SS = 4$$

$$EF_E = ES_E + D_E = 9 + 3 = 12$$

Activity F

$$ES_F = \max \begin{bmatrix} ES_C + SS_{CF} \\ EF_D \end{bmatrix} =$$

$$= \max \begin{bmatrix} 2 + 3 \\ 8 \end{bmatrix} = 8 \qquad \leftarrow FS = 0 \text{ and } SS = 3$$

$$EF_F = ES_F + D_F = 8 + 8 = 16$$

Activity G $ES_G = EF_E + FF_{EG} - D_G = 12 + 3 - 10 = 5$ $\leftarrow FF = 3$

$EF_G = ES_G + D_G = 5 + 10 = 15$

Activity H $ES_H = EF_F = 16$

$EF_H = ES_H + D_H = 16 + 2 = 18$

Activity I $ES_I = \max \begin{bmatrix} EF_G \\ EF_H \end{bmatrix} = \max \begin{bmatrix} 15 \\ 18 \end{bmatrix} = 18$

$EF_I = ES_I + D_I = 18 + 5 = 23$

Project duration, $T_S = EF_{\text{last activity}} = EF_I = 23$

BACKWARD PASS

The backward pass calculations are explained step by step as follows, based on the assumption that the project terminal time is equal to the project duration calculated in the forward pass as the early finish time of the last activity, $EF_I = 23$ weeks (assumed time units).

Activity I $LF_I = EF_I = 23$

$LS_I = LF_I - D_I = 23 - 5 = 18$

Activity float $= LF_I - EF_I = 23 - 23 = 0$

Activity H $LF_H = LS_I = 18$

$LS_H = LF_H - D_H = 18 - 2 = 16$

Activity float $= LF_H - EF_H = 18 - 18 = 0$

Activity G $LF_G = LS_I = 18$

$LS_G = LF_G - D_G = 18 - 10 = 8$

Activity float $= LF_G - EF_G = 18 - 15 = 3$

Activity F $LF_F = LS_H = 16$

$LS_F = LF_F - D_F = 16 - 8 = 8$

Activity float $= LF_F - EF_F = 16 - 16 = 0$

Activity E $LF_E = LF_G - FF_{EG} = 18 - 3 = 15$ $\leftarrow FF = 3$

$LS_E = LF_E - D_E = 15 - 3 = 12$

Activity float $= 15 - 12 = 3$

Activity D $\quad LF_D = LS_F = 8$

$\quad\quad\quad\quad LS_D = LF_D - D_D = 8 - 4 = 4$

$\quad\quad\quad\quad$ Activity float $= 8 - 8 = 0$

Activity C $\quad LF_C = \min \begin{bmatrix} LS_E - SS_{CE} + D_C \\ LS_F - SS_{CF} + D_C \end{bmatrix} =$

$\quad\quad\quad\quad\quad\quad = \min \begin{bmatrix} 12 - 4 + 6 \\ 8 - 3 + 6 \end{bmatrix} = 11 \quad\quad \leftarrow SS = 4 \text{ and } SS = 3$

$\quad\quad\quad\quad LS_C = LF_C - D_C = 11 - 6 = 5$

or $\quad\quad\quad LS_C = \min \begin{bmatrix} LS_E - SS_{CE} \\ LS_F - SS_{CF} \end{bmatrix} = 5$

$\quad\quad\quad\quad$ Activity float $= 11 - 8 = 3$

Activity B $\quad LF_B = LS_E - FS_{BE} = 12 - 2 = 10 \quad\quad\quad\quad \leftarrow FS = 2$

$\quad\quad\quad\quad LS_B = LF_B - D_B = 10 - 5 = 5$

$\quad\quad\quad\quad$ Activity float $= 10 - 7 = 3$

Activity A $\quad LF_A = \min \begin{bmatrix} LS_B \\ LS_C \\ LF_D - FF_{AD} \end{bmatrix} = \min \begin{bmatrix} 5 \\ 5 \\ 8 - 6 \end{bmatrix} = 2 \quad \leftarrow FF = 6$

$\quad\quad\quad\quad LS_A = LF_A - D_A = 2 - 2 = 0$

$\quad\quad\quad\quad$ Activity float $= 2 - 2 = 0$

The results of the forward and backward pass for the example project are presented in Table 5.1.

5.3   PRECEDENCE NETWORKING—DIAGRAM SOLUTION METHOD

A second solution method is presented here that is more visually oriented. Before explaining this method, it may be helpful to understand the different ways people learn. These are called the **styles of learning,** and people are often classified into one of three learner categories: auditory, visual, or kinesthetic. *Auditory learners* relate most effectively to the spoken word. They will tend to listen to a lecture and then take notes afterward, or rely on printed notes. *Visual learners* relate most effectively to written information, notes, diagrams, and pictures. *Kinesthetic learners* learn effectively through touch, movement, and space, and learn skills by imitation and practice. Few people use any one of these three styles exclusively; there is usually

TABLE 5.1 Forward and Backward Pass Results
PN Advanced Relationships

ACTIVITIES	EARLY START	EARLY FINISH	LATE START	LATE FINISH	FLOAT
A	0	2	0	2	0
B	2	7	5	10	3
C	2	8	5	11	3
D	4	8	4	8	0
E	9	12	12	15	3
F	8	16	8	16	0
G	5	15	8	18	3
H	16	18	16	18	0
I	18	23	18	23	0

Project duration = 23

some overlap in learning styles. For some people, the equation-based solution method presented in the previous section is easily learned. Others may find a more visually oriented solution technique easier to learn, thus the reasoning for this section.

The **PN diagram solution method** uses the step-by-step process of the forward and backward pass explained in the previous section strategically placed on the PN diagram (Figure 5.5). The early and late start and finish times for each project activity are determined and recorded literally on the PN diagram as the solution progresses. For illustration purposes, it is satisfactory to show several, but not all, of the steps of the forward and backward pass of the PN diagram. Let's begin the forward pass with activities A, B, C, and E. Referring to Figure 5.6, the steps taken are:

1. Given the early finish of activity A ($EF_A = 2$), the early start of activities B and C is 2 (that is, ES_B and $ES_C = EF_A = 2$).
2. The early finish of B and C is determined by adding the duration of each activity to the early start of that activity (that is, $EF_B = ES_B + D_B = 2 + 5 = 7$).
3. The early start of activity E is the maximum of the start values determined from activities B and C following the equation:

$$ES_E = \max \begin{bmatrix} EF_B + FS_{BE} \\ ES_C + SS_{CE} \end{bmatrix} = \max \begin{bmatrix} 7 + 2 = 9 \\ 2 + 4 = 6 \end{bmatrix} = 9$$

The last step in the series above demonstrates the usefulness of this solution method. While the equation alone can be used to determine a solution for the early start of activity E, the ability to visualize the solution as shown in Figure 5.6 can be very beneficial.

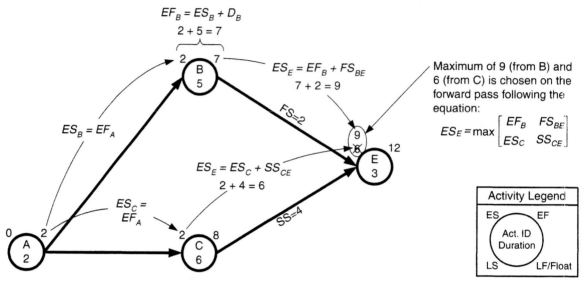

FIGURE 5.6   PN diagram solution technique—forward pass

Several steps in the backward pass process for the same project are shown in Figure 5.7. These steps include activities C, E, F, and G, with the backward action from E and F to C probably being the most difficult. The steps are as follows:

1. The late finishes of activities G and H are equal to the late start of activity I (LF_H and $LF_G = LS_I = 18$).
2. The late starts of activities G and H are determined by subtracting the duration of each activity from the late finish of that activity (that is, $LS_G = LF_G - D_G = 18 - 10 = 8$).
3. The late finish of F is equal to the late start of H ($LF_F = LS_H = 16$).
4. The late finish of E is determined by subtracting the finish-to-finish lag time from the late finish of G ($LF_E = LF_G - FF_{EG} = 18 - 3 = 15$).
5. As in step 2 above, the late starts of E and F are determined by subtracting the duration of each activity from the late finish of that activity.
6. The late finish of activity C is the minimum of the late finish values determined from activities E and F following the equation:

$$LF_C = \min \begin{bmatrix} LS_E - SS_{CE} + D_C \\ LS_F - SS_{CF} + D_C \end{bmatrix} = \min \begin{bmatrix} 12 - 4 + 6 = 14 \\ 8 - 3 + 6 = 11 \end{bmatrix} = 11$$

7. The late start of C is determined by subtracting the activity duration from the late finish of C ($LS_C = LF_C - D_C = 11 - 6 = 5$) or by taking

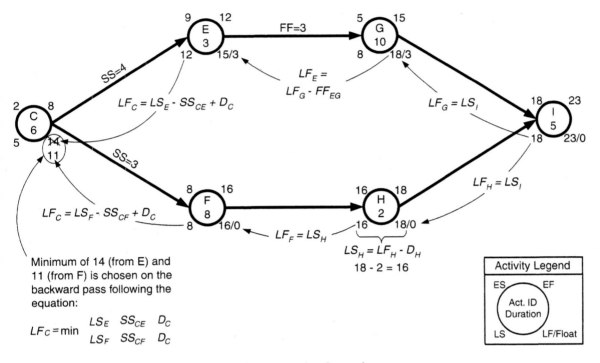

FIGURE 5.7 PN diagram solution technique—backward pass

the minimum of the late start values determined from activities E and F following the equation:

$$LS_C = \min \begin{bmatrix} LS_E - SS_{CE} \\ LS_F - SS_{CF} \end{bmatrix} = 5$$

5.4 PLANNED VERSUS ACTUAL PROJECT SCHEDULE

Both CPM and PN scheduling computations are performed under the assumption that all activity relationships will be executed as the model indicates. If problems (i.e., activity delays) arise in the actual schedule, the resulting adjustments in the CPM schedule can possibly lengthen the critical path but will have little effect on the individual relationships between activities. However, delays in the PN schedule may create more complicated problems. For example, let's assume that a start-to-start relationship has been assigned between two activities with a lag of 4 days, as shown in Figure 5.8. If activity A (*Install and finish drywall*) is delayed for 2 days for one of many reasons, the schedule still has activity B (*Paint facility interior*)

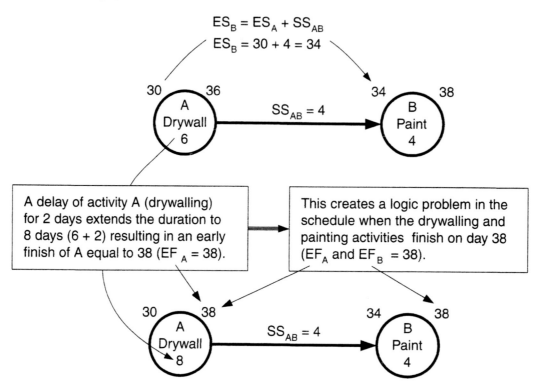

$$ES_B = ES_A + SS_{AB}$$
$$ES_B = 30 + 4 = 34$$

30 — 36 A Drywall 6 $SS_{AB} = 4$ 34 — 38 B Paint 4

A delay of activity A (drywalling) for 2 days extends the duration to 8 days (6 + 2) resulting in an early finish of A equal to 38 ($EF_A = 38$).

This creates a logic problem in the schedule when the drywalling and painting activities finish on day 38 (EF_A and $EF_B = 38$).

30 — 38 A Drywall 8 $SS_{AB} = 4$ 34 — 38 B Paint 4

FIGURE 5.8   Planned versus actual PN scheduling error

starting 4 days after the start of activity A (that is, $SS_{AB} = 4$ days). While this is still valid, the delay pushes the early finish of activity A to day 38 (36 + 2-day delay). By maintaining the SS relationship in the schedule, both activities A and B will finish on day 38, which is not a logical sequence. Both drywall finishing and painting cannot occur at the same time for any room in the facility. Obviously, the use of PN relationships requires careful comparison of the scheduled project to the actual project, with necessary adjustments being made to the schedule (i.e., model) routinely.

5.5   PROJECT SCHEDULING COMPUTER APPLICATIONS

The purpose of this section of the book is to provide a concise yet thorough overview of the use of project management software for construction scheduling. Manually going through the critical path method in Chapter 4 and precedence networking in this chapter provides a conceptual framework of the CPM and PN processes. Once the scheduling concepts of CPM and PN have been learned, it is helpful to see how these processes are utilized in the construction industry. Construction project planners have a strong interest in using computer software to manage and organize project work. Computer-based scheduling provides the primary advantage of immediate and accurate solutions of mathematical calculations where the accuracy of the schedule output depends solely on the accuracy of the input data.

The mathematical calculations in project scheduling are almost exclusively performed using some type of scheduling software such as Primavera's Project Planner (P3) or SureTrak (Primavera Systems, Inc., Bala Cynwyd, Pennsylvania) or Microsoft Project (Microsoft Corporation, Redmond, Washington). As stated in previous chapters, a recent survey of over forty project management professionals in the construction industry indicated that nearly 100 percent use project management software (Liberatore, Pollack-Johnson, and Smith, 2001). Results from that same survey showed that more than 50 percent of the construction professionals surveyed used a Primavera product, typically P3, for project management. Primavera's Project Planner is a relatively expensive, full-featured software package, while SureTrak is an inexpensive alternative that provides many of the features of P3 at reduced cost. Microsoft Project, while similarly priced to SureTrak, is directed more toward a mass market. In this text, SureTrak Project Manager (version 3.0) software was chosen due to its relative ease of use and the dominance of Primavera's project management products in the construction industry.

Upon completion of this section of the chapter, within SureTrak, you should be able to:

❏ Add or create a new project
❏ Modify project calendars
❏ Add and edit activities
❏ Assign relationships between activities
❏ Calculate the schedule

The assumption is made here that SureTrak 3.0 has been installed on a computer that at least meets the minimum system requirements as specified by Primavera.

This section of the text provides a brief overview of computerized project scheduling. Detailed coverage of Primavera software applied in the construction industry is available in a number of texts. The following two are recommended and are listed in the References section at the end of this chapter. For Project Planner (P3), refer to *Construction Scheduling with Primavera Project Planner* by Leslie Feigenbaum (2001), and for SureTrak, refer to *Scheduling with SureTrak* by David Marchman (2000).

5.5.1 CREATING A NEW PROJECT

The opening screen of SureTrak (Figure 5.9) gives access to the software tutorial, Project KickStart Wizard, project templates, and an existing or the last open project. In this case, a new project is being created without the use of the wizard or a template;

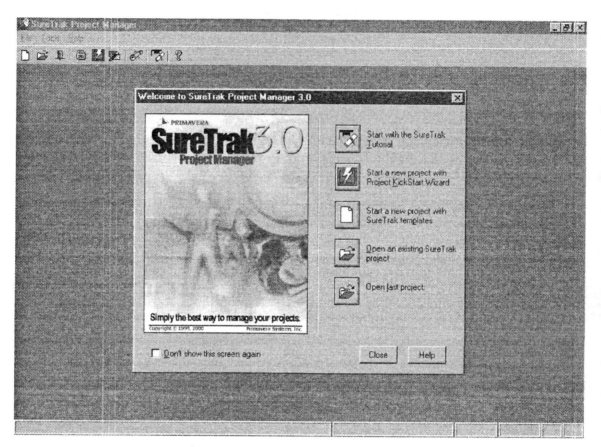

FIGURE 5.9   SureTrak entrance screen

therefore, choosing the Close button at the bottom of the box closes this option. A relatively blank gray screen appears except for the menu bar and icons at the top left of the screen. From left to right, the first three icons represent the file management functions within the software, specifically (1) New Project—this adds or creates a new project; (2) Open Project—this opens an existing project previously created; and Exit SureTrak—this closes all open projects and exits the software.

To create (add) a new project, click on the farthest left New Project icon. The SureTrak Project Manager dialog box will appear asking you if you want to run the software Wizard to assist in setting up the new project. In order to gain a better understanding of the basic capabilities of this software, click No to this question and the New Project dialog box appears (Figure 5.10). There are several fields to be completed here, but the intent is to provide information for those fields

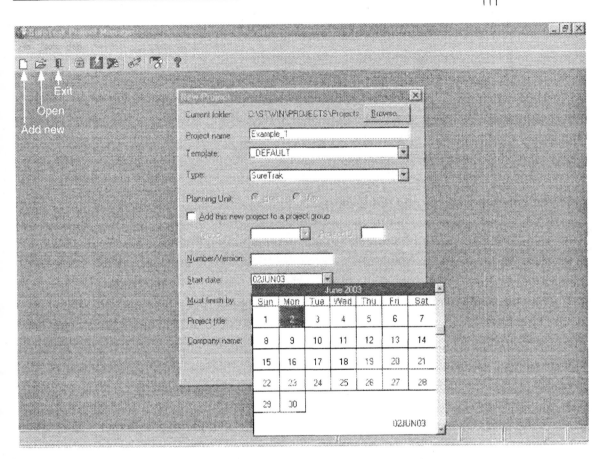

FIGURE 5.10 New project dialog box

that are necessary to create a relatively simple project. Of primary importance are the following:

❑ **Project name** This is a brief identification for the project and is equivalent to a file name for sorting and searching purposes. It is best to keep this name relatively short using alphanumeric characters and no spaces. SureTrak creates fifteen or more files for each new project with each file identified, in part, by this project name. For this example, the project name of "Example_1" was used.

❑ **Type** For a stand-alone project as being created here, this field should read SureTrak.

❑ **Start date** If an actual project start date is known, it should be entered in this field. If it is not known, some future date should be entered. This is easily modified later when an actual date is determined. In either case, avoid using the current date (the default field entry) as the project start date. If the current date is used and the project opens at a later date, critical activities can be shown as being behind schedule even though the project hasn't actually started. This causes inaccuracies in the critical activities and activity float calculations. A pull-down monthly calendar is accessible by clicking the down arrow to the right of the field bar (as shown in Figure 5.10). The calendar is advanced by clicking the down arrow on the slide bar (right side of the calendar). The date is chosen by clicking on it—Monday, June 2, 2003, is designated as this project's start date.

❑ **Must finish by**   At this point in the development of the project, leave this field blank. It is best to let the software determine the project finish date through schedule computations rather than attempting to provide a date. If a date is entered that does not correspond to the computed finish date of the latest project activity, the critical activities and activity float calculations are adversely affected.

❑ **Project title**   This title is typically longer than the project name (i.e., file name) and defaults as the project title printed on hard copy for tabular and graphic reports.

❑ **Company name**   This name is, by default, also printed on the hard copy for tabular and graphic reports. The project planner's name can also be inserted here.

Once all the pertinent information is entered in the New Project dialog box, click OK and the new project schedule is created. A new screen appears resembling an electronic bar chart, with the activity descriptions and other information (i.e., I.D., duration, etc.) on approximately the left one-third of the screen and a Timescale and activity bars on the remainder of the screen. (Figure 5.11)

A portion of the project schedule for the three-unit townhouse is shown in Figure 5.12. The schedule portion is used in this section of the book to illustrate the creation of a project schedule within SureTrak. Note that only seven of the projects' activities are shown here; for the entire schedule, refer to Appendix A. In this section of the schedule (primarily framing and rough-ins), advanced PN activity relationships are present and are shown on the relationship arrows in Figure 5.12.

5.5.2   MODIFYING TIMESCALE AND PROJECT CALENDARS

The starting date of the project corresponds to the *data date* in the software as June 02, 2003 (Figure 5.11). The data date is the "as of" date within the project, and SureTrak uses the current data date to calculate the schedule. When creating a new project, the default data date is the specified project start date. By clicking the left directional arrow at the bottom of the right side of the screen (activity bar and Timescale section), a red line (indicating the project *data date*) can be seen at the beginning of the 06/02/03 week. As a project

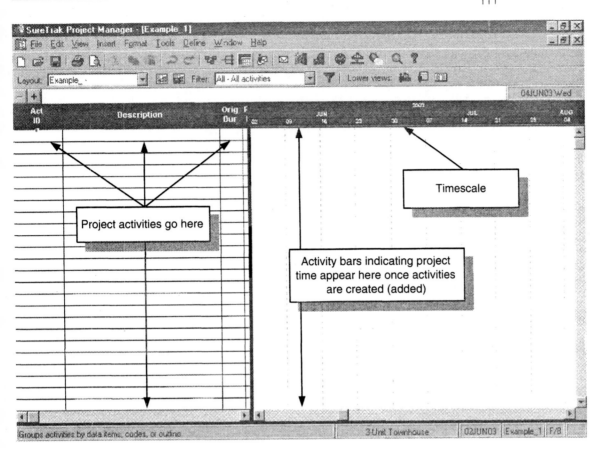

FIGURE 5.11   Default screen for example project

proceeds, the data date can be moved forward toward the end of the project. The data date can be updated by dragging the data date line across the bar chart using the Project Spotlight feature or by entering the data date in the Progress dialog box.

The Timescale or calendar dates for the project are shown in the right upper portion of the bar chart in Figure 5.11. Below the Timescale portion of the bar chart, the project activity bars on the right of the screen appear as the activities are defined in the activity description area to the left. Until relationships are assigned between activities, each activity will begin on the project start date (and data date). At this point, the Timescale calendar in Figure 5.11 shows the Monday of each week. For the example project with time units in days, it is more appropriate for the Timescale to show days rather than weeks prior to defining activities. The Timescale is modified via the Timescale window or menu box. This is accessed by double-clicking on the month or year of the actual Timescale or selecting Format, Timescale from the pull-down menu at the top left of the

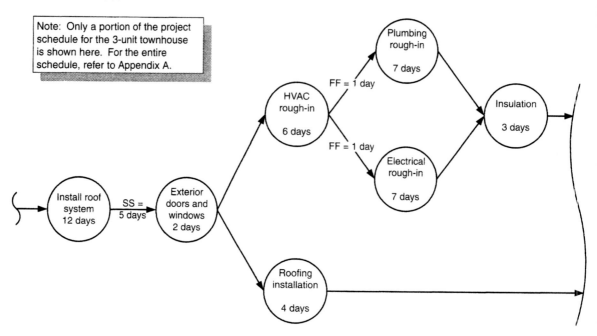

FIGURE 5.12   Example schedule—portion of the three-unit townhouse

screen (Figure 5.13). Important settings of the Timescale window include the following:

❑ **Density**   This is the level of compression for the calendar dates (or ordinal dates—see the explanation on the next page) on the Timescale. In other words, for any given time unit (i.e., hour, day, week, month), the density setting defines the amount of space per unit time or the amount of space a time unit occupies on the Timescale. A more dense setting shows more calendar days with shorter activity bars, while a less dense Timescale expands the activity bars with fewer calendar days shown. For the example project, increase the density to the maximum amount by moving the slide button (i.e., blinking box) to the farthest left position on the density bar.

❑ **Begin Date and End Date**   These dates set the time span viewed (and printed) on the bar chart. The default *rolling dates* of -7d for the Begin date and 21d for the End date automatically shift the Timescale of the bar chart as a project progresses—the default shift is 7 days (1 week) prior to the project start or data date and 21 days (3 weeks) after the project completion date. SureTrak allows four options instead of rolling dates, namely Calendar date, Start date, Data date, and Finish date, which are accessed by clicking on the SD+ and FD+ buttons. The Begin and End dates can be tied to any of these, or rolling dates can be used. Set the Begin date to -2d and the End date to 2d for the example project.

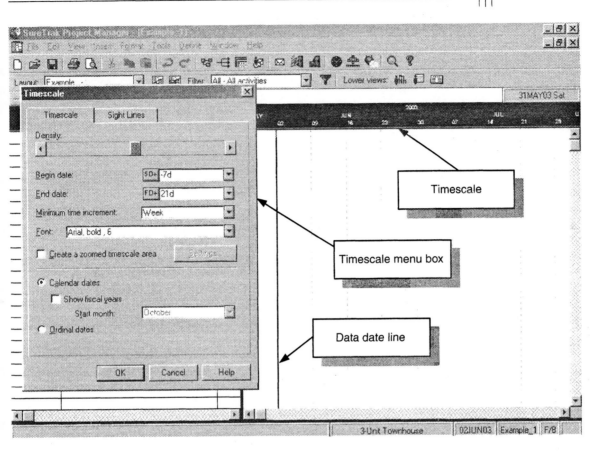

FIGURE 5.13 Timescale menu box

❏ **Minimum Time Increment** The default time unit is week. The
SureTrak options here are hour, day, week, month, quarter, and year.
For the example project, choose day as the time unit. This also can be
changed by double-clicking on the bottom portion of the Timescale
where the Monday dates are shown; each double click cycles the time
unit to the next smaller value.

❏ **Calendar Dates and Ordinal Dates** Project schedules are commonly
shown in calendar dates that include weekend dates. Ordinal dates or
numerical dates (e.g., day 1, 2, 3, etc.) are simpler for the beginning
project planner. The example schedules solved manually throughout this
book have used ordinal dates to this point. However, when the project
is executed, ordinal dates hold little meaning to most people. Most
commonly, calendar days are utilized in a project. This is the reason that
the SureTrak default uses calendar dates. Therefore, for our example
project, the first ordinal date is the calendar date of June 02, 2003.

Prior to defining project activities, a project calendar must be chosen or defined. Calendars define when activity work is scheduled during the workweek. This includes the days of the week and the hours of the day when work is scheduled. Calendars also define the nonwork exceptions to the workweek (i.e., weekends, holidays, and nonworking hours of the day). The default project calendar in SureTrak is called a Global calendar. Two other choices are available to the user: the Normal workweek calendar and the Seven 24-hour days calendar. The Global calendar defines the normal company work and nonwork dates and times. This calender also defines the number of work-hours per day and the number of workdays per week for the entire project. Those values are used in SureTrak to calculate the number of days and weeks represented by a specified number of hours of activity duration.

The Project Calendar window (Figure 5.14) is accessed by selecting Define and then Calendars from the pull-down menu at the top of the screen. The Project Calendar window provides the user the ability to select one of three calendar types, to modify an existing calendar, or to create a new calendar. SureTrak supports multiple calendars within any given project; specific calendars can be assigned to specific activities. This feature allows activities to be performed during different-length workdays or workweeks. Working and nonworking days as well as working and nonworking hours are shown on the default Global calendar (annotated in Figure 5.14). The normal workweek for the Global calendar is Monday through Friday, with the weekend as nonworking days; and the workday hours are 8:00 A.M. to 5:00 P.M., with 1 hour off for lunch (12:00 to 1:00 P.M.). The default workweek and workdays of the Global calendar are used for the example project given below.

5.5.3 ADDING ACTIVITIES, DEFINING RELATIONSHIPS, AND CALCULATING THE SCHEDULE

Figure 5.12 is a logic diagram of the example project used in this section of the book to illustrate the creation of a project schedule within SureTrak. The example project is a portion of the project schedule for the three-unit townhouse.

To add an activity to the project, three primary types of information must be completed as follows:

❑ **Act ID**   The activity ID, or identification, is a set of alphanumeric characters (letters and/or numbers) that gives a unique identity to individual activities. SureTrak uses an autosequencing feature that numbers each new activity in increments of 10, starting at 1000. By simply clicking on the Act ID field, a new activity is added to the project. SureTrak tracks project activities by the Act ID. Therefore, it is advisable to use the same number and similar sequence of characters for activities so that activities can be tracked on a similar basis. For the example project, the default sequence (i.e., 1000, 1010, 1020, . . .) will be used. After checking the default value in the Act ID field, click on the Description field.

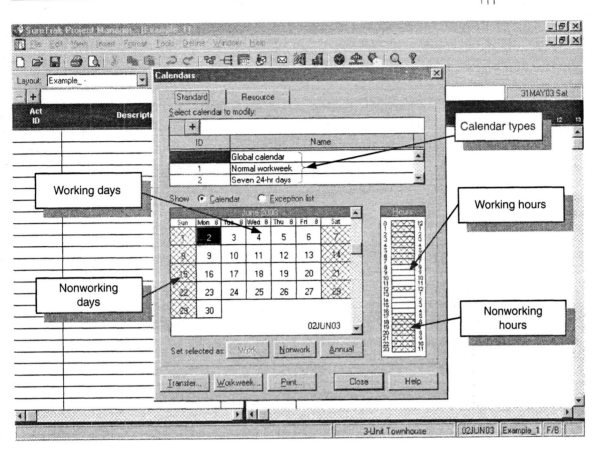

FIGURE 5.14   Project calendar window

❑ **Description**   This alphanumeric field is used to give a brief description of the activity. The description is normally two to three words that provide sufficient information to communicate the specific activity to various people involved in the project. After completing the activity description, click on the Orig Dur (original duration) field.

❑ **Orig Dur**   The original duration is the time assigned to the activity to complete or accomplish the work. The activity duration was defined earlier in Chapter 3 as the time it takes to complete the activity from its start to finish. This time can be represented in minutes, hours, days, weeks, or other units. The time unit of days is often used in construction scheduling because many construction activities require multiple days (2 or more) to complete. The default time unit in SureTrak is days as indicated by the "1d" in the Orig Dur field when activities are first defined.

Activities should be entered into the bar chart for the example project (Figure 5.12), as illustrated in Figure 5.15. Note that all defined activities start on the project start date or data date. This occurs because relationships have not been defined between activities. The screen view of the activity bars defaults to a red-colored bar for critical activities and a green one for noncritical activities. At this point, activity 1000—*Install roof system*, with a 12-day duration—is the only critical (red bar) activity. These are actual screen colors although they are visible only as gray-scaled shading in Figure 5.15.

In Figure 5.15, activity 1000 has a start date at the beginning of the day on Monday, June 02, 2003 (i.e., project start date), and a finish date at the end of the day on Tuesday, June 17, 2003. This is a total of 16 days, but only 12 of those days are working days (Monday through Friday for the default Global calendar). This includes two weekends (Saturday and Sunday), or a total of 4 nonworking days (June 7, 8, 14, and 15). Similar behavior can be observed for activities 1020, 1030, and 1040, because each of these has a duration greater than 5 days,

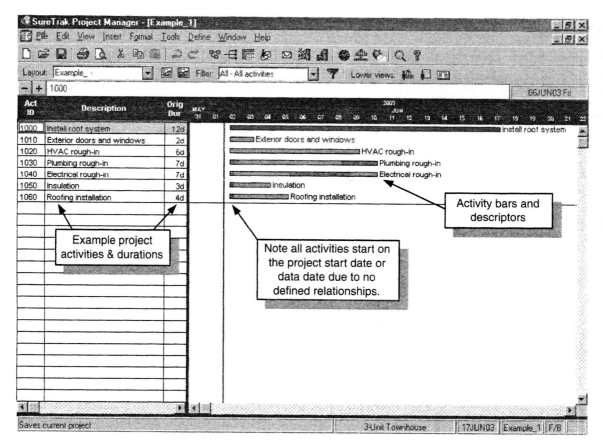

FIGURE 5.15   Bar chart—example project activities

scheduling them past the first and into the second workweek with a break in the activities during the first weekend. Most of these initial scheduled times will change with the inclusion of activity relationships.

Activity relationship can now be defined. There are several ways that Sure-Trak allows the inclusion of relationships between activities. One of the simplest ways is to move the cursor arrow to either the start or finish of a predecessor activity bar. The arrow is placed at the start of an activity to insert a start-to-start relationship and at the finish of an activity to insert either a finish-to-start or a finish-to-finish relationship. When the arrow approaches the start or finish of the activity bar, a pitchfork-shaped icon appears (Figure 5.16). This icon is dragged (click and hold the left mouse button) from the predecessor activity to one of its successor activities.

For example, the relationship between activity 1000—*Install roof system*—and activity 1010—*Exterior doors and windows*—is a start-to-start relationship with a lag of 5 days (SS = 5 days). The pitchfork icon is activated at the start of predecessor activity 1000 and dragged to the start of the successor activity 1010 and released. After releasing, the Successor Activity window appears for the user to complete the details of the relationship (Figure 5.17). In this case, the lag must be increased from 0 to a value of 5. This process is continued until all relationships have been inserted between all activities, resulting in the completed bar chart schedule (Figure 5.18) and the precedence diagram schedule (Figure 5.19).

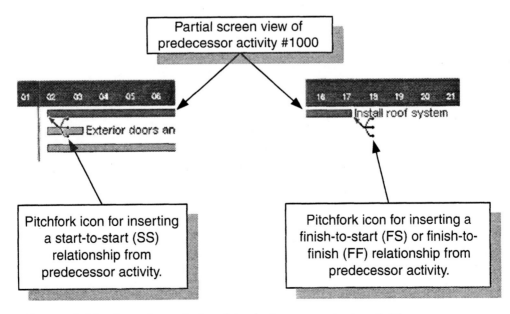

FIGURE 5.16   Inserting relationships between project activities

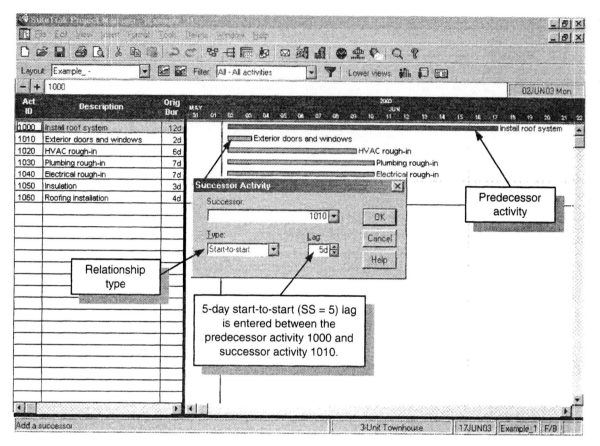

FIGURE 5.17   Successor activity window

The precedence diagram is called a PERT chart in SureTrak and other software. While this is an accepted standard in project management software, it is a misnomer, as previously explained in Chapter 4. PERT, or program evaluation and review technique, is a probability modeling method for scheduling projects that have highly variable activity durations. PERT is applied using AON diagramming, and this probably contributes to the misunderstanding. However, most scheduling software packages do not have PERT capabilities in regard to variable activity durations, and a more appropriate name to use is AON, sequence, or precedence diagram rather than PERT diagram.

Notice that as a relationship is inserted between any pair of activities in SureTrak, the schedule is automatically computed for all activities. SureTrak, by default, calculates the entire schedule every time you change anything that might affect it (i.e., relationship type, lag, or activity duration). You can turn off this automatic calculation by choosing Tools from the pull-down menu and selecting Schedule.

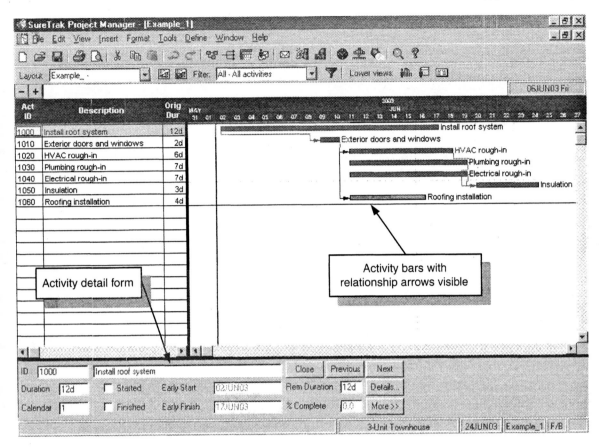

FIGURE 5.18   Final bar chart—example project

5.6   SUMMARY

The precedence network, often referred to as the precedence diagramming method (PDM), is a CPM scheduling process whereby project activities are graphically represented as nodes and activity relationships as arrows. Precedence networking (PN) differs from CPM in that activity relationships can be represented in a more complex form. The most common PN constraints include the traditional *finish-to-start* (FS) relationship used in CPM and the *start-to-start* (SS) and the *finish-to-finish* (FF) relationships. Not only does PN permit these alternative activity relationships, but it also allows for time-lag quantities to be assigned to relationships.

The FS relationship, where one activity must finish before the start of its successor, is the most common relationship in construction scheduling and is predominately used to represent activity sequence. The SS relationship is

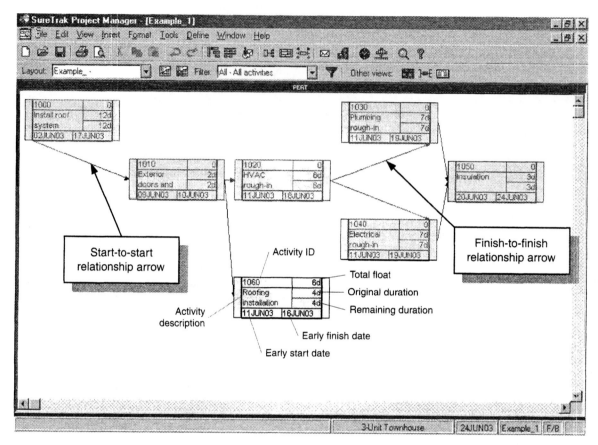

FIGURE 5.19   Final precedence diagram (PERT Chart)—example project

applied to schedule the start of a successor activity at the same time as or after some lag time to the start of the predecessor activity. The FF relationship places no restriction on activity start times, but rather requires that the finish time of the successor activity be no earlier than or after some lag time before the finish time of the predecessor activity. A lag quantity is a time period that separates or designates a waiting period between two activities. The time lag is associated with the relationship arrow between two activities—a predecessor and successor.

The computational method used in precedence networking is more complex than the previous CPM method in which only FS activity relationships were allowed. Advanced activity relationships (i.e., SS and FF) complicate the forward and backward pass. On the forward pass, the SS relationship is of particular concern due to the connection of activity start times rather than the simple transferring of the early finish time of a predecessor activity to the early start time of a successor activity. Similarly, on the backward pass the FF relationship is troublesome. Detailed example computations, figures, and tables help explain this procedure.

The PN diagram solution method is an alternative solution method for the more visually oriented learners. This method uses the step-by-step process of the forward and backward pass strategically placed on the PN diagram. Activity times are determined and recorded literally on the PN diagram as the solution progresses.

Construction project planners have a strong interest in using computer software to manage and organize project work. Computer-based scheduling provides the primary advantage of immediate and accurate solutions of mathematical calculations where the accuracy of the schedule output depends solely on the accuracy of the input data.

CHAPTER 5 QUESTIONS/PROBLEMS

1. *Team Activity*-Create and sketch an AON precedence diagram of a realistic project of at least twelve activities, including at least three advanced activity relationships (i.e., FF and SS). At least three activities must have two or more predecessors. Use lettering (i.e., A, B, C, ...) to designate project activities. Assign accurate activity durations for each project activity. Draw an AON network diagram of the project indicating all relationships, early and late times, and float times with appropriate coding, and compute the critical path for this project using the forward and backward pass method.

2. *Precedence Networking*-All the precedence relationships for the following project are direct finish-to-start relationships with no lags except for the following:

C to D	Start-to-start with a lag of 1
D to E	Finish-to-finish with a lag of 3
A to F	Start-to-start with a lag of 2
H to I	Finish-to-finish with a lag of 4
L to M	Start-to-start with a lag of 1

Accurately draw an AON network diagram indicating all relationships, early and late times, and float times with appropriate coding, and compute the critical path for this project using the forward and backward pass method.

3. For the AON diagram on the next page, determine the early and late times, float times, and project duration using the PN forward and backward pass method.

ACTIVITY	PREDECESSORS	DURATION, DAYS
A	—	5
B	A	6
C	B	3
D	C	4
E	D, G	5
F	A	8
G	F, J	3
H	—	3
I	H	2
J	I	7
K	F, J	2
L	H	7
M	L	4
N	K, M	3

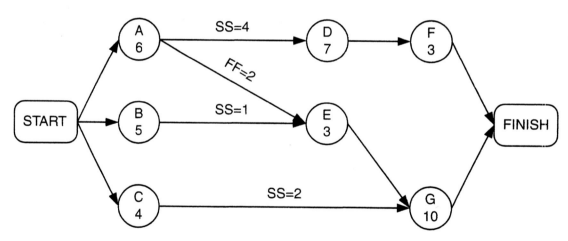

4. *Computer Application—House Construction Project*—Using Primavera
   SureTrak software, generate both a bar chart and an AON (SureTrak's
   PERT) diagram of the house construction project given in the table on the
   next page. Using the default calendar in the software, begin the project on
   a Monday during one of the future summer months (i.e., current or next
   year). Size the bar chart and AON diagram to appropriately fit within two
   pages each (8.5-in. by 11-in. page size), and print both the bar chart and
   AON diagram (total of four pages). Cut and paste the chart and diagram
   to show one continuous drawing for each.

House Construction Project			
JOB NAME	**DESCRIPTION**	**IMMEDIATE PREDECESSORS**	**TIME (DAYS)**
A	Excavate and pour footers	—	4
B	Pour concrete foundations	A	2
C	Erect frame and roof	B	5
D	Lay brickwork	C	6
E	Install drains	B	1
F	Pour basement floor	E	2
G	Install rough plumbing	E	3
H	Install rough wiring	C	2
I	Install air conditioning	C, F	4
J	Fasten plasterboard	G, H, I	10
K	Lay finished flooring	J	3
L	Install kitchen equipment	K	1
M	Install finished plumbing	K	2
N	Finish carpentry	K	3
O	Finish roofing and flashing	D	2
P	Fasten gutters and downspouts	O	1
Q	Lay storm drains	B	1
R	Sand and varnish floors	N	2
S	Paint	L, M	3
T	Finish electrical work	S	2
U	Finish grading	P, Q	2
V	Pour walks, and landscape	U	5

REFERENCES

See References on page 199.

C H A P T E R

6

RESOURCE SCHEDULING, LEVELING, AND ALLOCATION

O B J E C T I V E S

This chapter provides knowledge in the areas of:

- ❑ Management of construction resources
- ❑ Resource scheduling—resource profiles
- ❑ Resource leveling with no resource limits—sum of squares method
- ❑ Resource allocation with fixed amounts of resources—series and parallel methods

6.0   OVERVIEW

This chapter addresses general resource management in construction projects. The chapter begins with a discussion about construction resources and then goes on to discuss construction resources in regard to resource management in the construction industry. The chapter explains the general trends of resource-constrained schedules as well as the analytical methods used for resource management—scheduling, leveling, and allocation. Information on resource scheduling is highlighted using resource loading and profile diagrams. Next, the chapter describes resource leveling with a comprehensive example of the sum of squares analytical method. Finally, the chapter focuses on the series and parallel methods of resource allocation and illustrates the series method through an example problem.

6.1   CONSTRUCTION RESOURCES

In Chapter 1, construction project management was defined as the planning, scheduling, and controlling of construction tasks or activities to accomplish specific objectives by effectively utilizing appropriate **resources** in a manner that minimizes costs and maximizes customer/owner satisfaction. Those resources include time, materials, labor, and equipment. Cash or project funding may also be included as a project resource. Resource management is one of the most important aspects of construction project management in today's economy because the construction industry is resource-intensive and the costs of construction resources have steadily risen over the last several decades.

Resources for a construction project are usually first determined in some detail by an estimator or estimating group or team. The resources are classified by type following various phases of the project. For example, large structural columns of a building typically require concrete placement (see Figure 6.1). From the dimensions of the columns, the amount, or volume, of concrete is determined in order to estimate the cost of this material for this particular step in the building process. It is important to understand that the concrete is only one of several resources needed for this step. The concrete forms, structural reinforcing steel (rebar), overhead crane and bucket, and human labor are also required for column layout, concrete form erection, placement of steel, concrete pour and cure, and form removal.

6.2   MANAGEMENT OF RESOURCES

Up to this point, the critical path and precedence networking methods have only been constrained by the resource of time and sequence relationships among activities. In other words, the only limiting factors for project activities after scheduling are the start and finish dates and times. Considering only the time resource and activity precedence, any given activity can start once the technological sequence of predecessors for that activity has been satisfied, and can finish after its designated duration. But project activities consume more resources than just time, as explained in previous chapters. In Chapter 3, the guidelines for estimating activity durations instruct the reader to assume that *materials, labor, equipment, and other resources* have unlimited availability for any given activity. The guidelines further state that activities should be considered individually; resource requirements of other preceding, succeeding, or concurrent activities are not to be considered, and conflicting demands among concurrent activities for resources should be ignored during the duration-estimate process.

FIGURE 6.1 Structural column concrete placement
SOURCE: *Construction Digest*

The assumption of unlimited and nonconflicting resources is necessary in scheduling as part of the activity-definition and duration-estimate processes. At these stages of the scheduling procedure, the potential resource conflicts among activities are not explicitly considered, and the decisions made for any given activity are usually independent of other scheduling activities. However, these are not isolated decisions and can be influenced by the consideration of concurrent activities and conflicting resources even by experienced project planners or estimators. It is best to remember that project scheduling that includes resource management is a multistep process of developing the time-constrained schedule and then considering resource conflicts or the resource-constrained schedule.

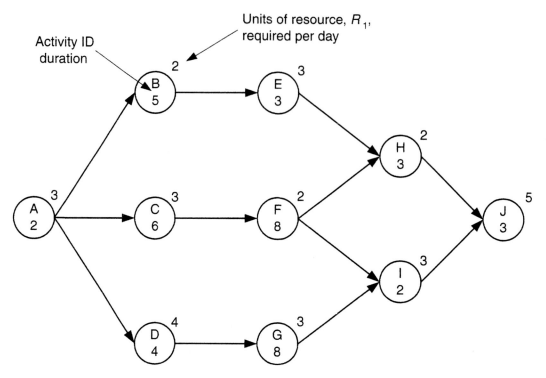

FIGURE 6.2 AON diagram—example project

Often the project planner utilizes the time-and-precedence–based schedule as a basis for the management of resources for the project. The schedule is developed considering only the constraint of time and precedence among activities (time constraints) resulting in early and late start and finish dates/times for all activities. The planner then concentrates on the resource requirements (resource constraints) that are estimated for each project activity, particularly equipment, materials, and labor.

In a resource-constrained schedule, resource units are limited at some maximum level that, for any given resource, is assumed to be less than required during one or more of the time periods within the project. For example, assume two activities are concurrent or parallel (occur in the same time period) in a time-constrained, early start schedule. Figure 6.2 illustrates this simple example project with an AON diagram. Figure 6.3 shows the bar chart of that same example. Assume for this example that activities C (critical) and B (noncritical) in Figure 6.3 each involve excavation and each require the only available excavator (a backhoe) for the project. The solution for this resource conflict is to either (1) delay the noncritical activity, B, until the critical activity, C, is completed and then shift the backhoe to the noncritical activity or (2) lease, rent, or purchase another backhoe for the second, less critical activity, B. It can be seen in Figure 6.3 that the noncritical activity, B, has sufficient float time (6 days) to postpone its early start (ES) to alle-

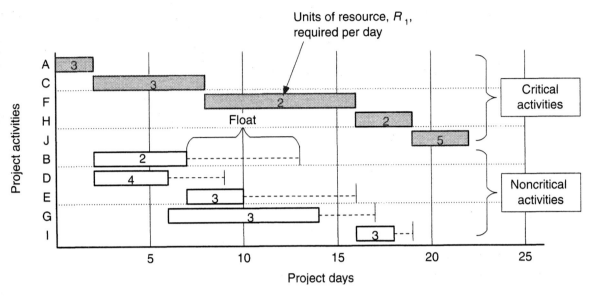

FIGURE 6.3 Early start schedule bar chart—example project

viate the resource conflict. This would probably be the most cost-effective solution. In other words, the noncritical activity would be delayed until the critical activity (i.e., float = 0) is completed, and then the noncritical activity would start. An important result of delaying the noncritical activity, B, to alleviate this resource conflict is that activity B is delayed until its latest late start (LS). By using the entire available float of activity B, the activity becomes critical although it is not on the original time-constrained critical path.

In the example above, the start/finish time of a noncritical activity in the resource-constrained schedule was modified in order to maintain the maximum resource limit for the excavator (a single backhoe). The resource management techniques used to modify the time-constrained schedule tend to result in one or more of the following general trends for the resource-constrained schedule:

❑ Project activity float time is reduced, consequently making noncritical activities more critical.

❑ Early and late start/finish schedules tend to be no longer unique (for noncritical activities).

❑ Float is a function of both precedence relationships and resource limitations.

❑ The time-constrained critical path(s) may be different in a resource-constrained schedule.

While not always the case, it's interesting to note that all four of these trends are true for the simple example described above.

In a resource-constrained project, the amount of resources needed during one or more time periods is limited to less than required by two or more activities occurring simultaneously. Resource management techniques must be employed to distribute resource usage period by period in order to minimize resource variations over time. Resource management can be divided into three main stages or steps for the analysis and solution of resource-constrained schedules as follows:

- ❏ Resource scheduling
- ❏ Resource leveling
- ❏ Resource allocation

Each of these serves a slightly different purpose and is applied in the order listed as constraints on project resources become more critical. *Resource scheduling* simply provides a profile of resource usage during the duration of the project. *Resource leveling* uses the float time of noncritical activities to redistribute activity start and finish dates in order to eliminate or at least reduce resource conflicts. Because resource leveling is only applied to noncritical activities, the critical path remains untouched and the project duration is unchanged. This technique is often applied when the project completion date is specified and fixed. In the third stage, *resource allocation*, limited resources are allocated to project activities by guidelines that assign resources to activities according to their level of criticality. This is the most common type of resource management technique due to the realities of resource availability and limitations. Resource allocation frequently extends the project duration, but the objective is to finish the project as close as possible to the time-constrained completion date.

The use of microcomputers and sophisticated project scheduling computer software is commonplace in the construction industry, especially on large projects (i.e., greater than 100 activities—see, for example, Figure 6.4). This is principally true for scheduling project activities and managing resources. The manual solution of resource management for most projects is very difficult and unlikely to be done in practice. However, just as with CPM and PN methods, using the manual solution to examine and understand the resource management techniques is valuable and necessary.

Before we provide a detailed explanation of the resource management stages, there is a very important assumption to clarify regarding resource requirements of individual activities prior to performing any type of resource analysis. Resource conflicts occur between concurrent activities, not within a single activity. For any given project activity, it is assumed that sufficient resources are available and fixed to accomplish the activity under normal work conditions during its specified duration. In other words, a certain resource load was assumed in order to estimate the duration of each activity. If the level of a particular required resource is not available for a given activity, that activity duration must be increased to accommodate the limit on that resource. For example, if six workers are needed for a concrete pour (single activity) but only four workers are available, the duration of the concrete pour must be adjusted longer to allow time for four workers to accomplish the work.

FIGURE 6.4 Large multiactivity bridge project in metropolitan setting
SOURCE: *Construction Digest*

6.3   RESOURCE SCHEDULING—RESOURCE LOADING AND PROFILES

Resource scheduling, often referred to as resource loading, is likely to be the most common resource management method used by project planners. The network model process and use of computerized scheduling provides the ability to organize resource information and present that information over the duration of a project. Resource requirements for each project activity must be known; this information has typically been established as a necessity in the estimating process.

To schedule resources, resource requirements are specified for each project activity and resource loading is shown in a graphical profile. This *resource profile* is determined by summing the resource requirements over time during the project using either an early or late start schedule, which was previously developed in the time-constrained schedule. The early start schedule is advantageous because it preserves the float time for all noncritical activities. Throughout the duration of the project, the total amount of the resources required is the sum of the resources required for the individual activities. Referring to the backhoe excavator example, two units of this resource were required at the same time because two activities occurring simultaneously required one unit each.

In the resource scheduling process, a resource profile is generated for a particular resource, with the time line of the project typically shown on the horizontal axis. For the example project represented in Figures 6.2 and 6.3, units of resource, R_1, required per day for each activity are shown numerically in both figures. Please note that the resource R_1 is no longer referring to the backhoe excavator example. Rather this resource could be assumed to be a labor or material requirement for each activity. The resource profile, commonly referred to as a *resource loading diagram*, is shown in Figure 6.5 accompanied by the project bar chart on a matching time line. It's evident in Figure 6.5 that the resource, R_1, is being utilized a maximum of 9 units at the beginning of the project to a minimum of 2 units for 3 days toward the end of the project.

The resource profile provides information on resource loading according to how resources were specified in the original estimates for project activities. In the development of the resource profile, unlimited resources are assumed. In other words, resource limits and thus resource conflicts are not specified. Let's assume that the daily totals of the resource, R_1, shown in Figure 6.5 are the number of general laborers required (i.e., labor requirements) for the example project. Again, the required number of laborers varies from a minimum of two to a maximum of nine over the 22-day project duration, creating the peaks and valleys in the profile. Managing the variation, or smoothing out the peaks and valleys, for this particular resource is a difficult task for the construction contractor on this project. If the contractor has multiple job sites, it may be possible to assign the specific number of laborers needed to maintain the existing early start schedule on this project, as long as this labor requirement can be matched to the needs of other job sites. This is only possible if the total number of laborers for all job sites is equal to the total number of laborers available. Otherwise, the distribution of laborers to different job sites requires careful prioritization. If the contractor has fewer or only one active job site, this would require the hiring and laying off of personnel on a daily basis. The negative psychological effect on workers and the cost of repeated hiring and firing of workers would be significant. An alternative, and surely more effective, solution is to utilize *resource leveling* to reduce the variations in the resource profile.

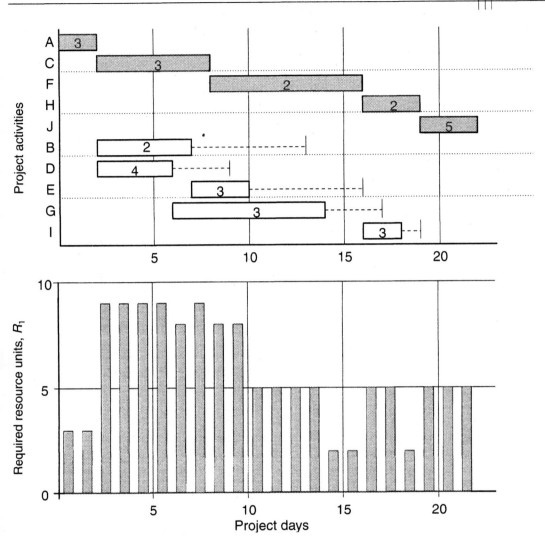

FIGURE 6.5 Resource scheduling—profile of example project

6.4 RESOURCE LEVELING (FIXED PROJECT DURATION)

Resource leveling is a broadly applied and common term in project resource management. The primary objective in resource leveling is to reduce the peaks and valleys in a resource profile without increasing the project duration. The technique basically involves delaying project activities in a time-constrained schedule so that the project resources are redistributed in order to eliminate resource conflicts. In the example of a backhoe conflict between two activities, the most cost-effective option involved delaying the noncritical activity. The other option involved spending the necessary money to make a second backhoe available. The latter technique is called a *time-cost trade-off* and is explained in Chapter 7.

In resource leveling, a resource profile is generated (typically from the early start schedule) and then the resource variations (i.e., peaks and valleys in the profile) are reduced by shifting noncritical activities (i.e., those with float time > 0) to non–peak time periods in the schedule. Activity float time is determined from the time-constrained, early start schedule under the assumption of no resource constraints. Resource leveling is done without delaying the activities along the critical path(s), and thus the scheduled finish time of the project is unchanged, or fixed. In other words, resource leveling does not increase the project duration. To accomplish this, the assumption of unlimited resources (resource limits not specified) is still in effect, as it was to generate the resource profile.

In resource leveling, the technique of utilizing the float time to shift noncritical activities in order to smooth out the profile seems relatively straightforward. But the application of the technique involves first selecting a combination of noncritical activities to delay and then determining the specific delay for each activity that smoothes out the profile best. A good example is the project depicted in Figure 6.5. Referring to Figure 6.5, it is somewhat obvious that shifting the requirements for general laborers (i.e., resource R_1) during days 3 through 10 to days 11 through 22 of the project can smooth the profile. Activities B, D, E, G, and I each are noncritical in the time-constrained, early start schedule and have available float time. To smooth the profile, activities B and E on one noncritical path and/or activities D and G on a different noncritical path could be delayed some or all of their float time (i.e., a total of 6 days for the B-E path and 3 days for the D-G path) without extending the project duration. Because path B-E has the largest float time, it may be the best to examine first. Delaying activities B and E to their latest start times (i.e., use all available float) results in the resource profile shown in Figure 6.6. The bar chart in Figure 6.6, shows that activities B and E, while noncritical in the time-constrained schedule, are now critical and make up a parallel critical path to the original schedule. The action of delaying activities B and E seems to have reduced the peaks and valleys in the original resource profile although this is difficult to discern visually.

In this example, the project was small—only 10 activities. The choices of noncritical activities to delay were few and relatively simple, and it was not difficult from the resulting resource profile to visually discern the reduction of resource variations. In this simple example, the use of simple guidelines such as "largest float time" or "minimum activity duration" is appropriate and provides a valuable aid to implement the resource leveling process. For larger and more complex projects (normal for construction scheduling), the task becomes more challenging, particularly when done manually. Given multiple activities and relationships, one must determine which noncritical activities to choose to delay first, which combination of activities to choose, the order in which they are cho-

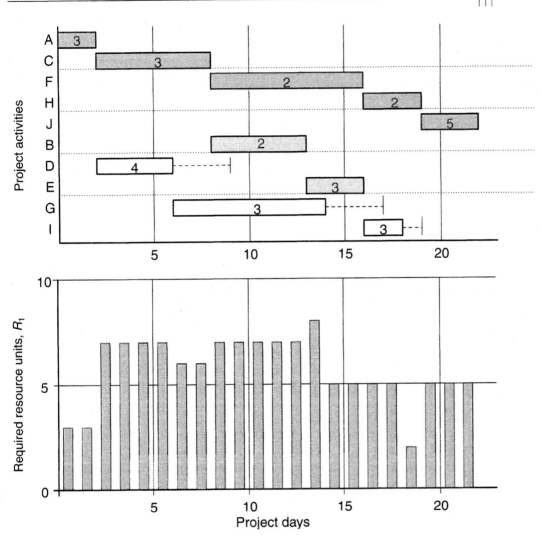

FIGURE 6.6   Resource leveling by delaying noncritical activities B and E

sen, and how much each activity would be delayed in order to reduce resource variation. The number of choices and outcomes for this type of complex problem increases very quickly.

The large number of possible outcomes for large, complex projects makes the resource leveling problem a combinatorial, nondeterministic, polynomial time complete problem (Savin, Alkass, and Fazio, 1996). Combinatorial problems are ones that have a large number of possible combinations of factors, and thus which result in a significantly large number of possible solutions. In the resource leveling problem, many combinations of activity start times (for noncritical activities) exist, with each combination representing a different project schedule. As project size (i.e., number of activities) increases, the number of combinations increases exponentially. For even relatively small projects (25 to 40 activities), the number of combinations is so large that developing a solution for all possible combinations is unmanageable. And as the number of activities increases, solving all combinations is difficult even with the use of computers.

Analytical approaches yielding deterministic solutions for these types of combinatorial problems are not successful. As an alternative, several heuristic-based solution techniques have been developed. A heuristic, as it is applied here, is a general "rule of thumb" or set of guidelines to direct the problem-solving process. A heuristic is designed, in this case, to reduce the number of combinations analyzed in order to reduce the amount of work required to get a "workable" solution. A workable solution is one that satisfies most of the requirements for a good solution but may not be the best solution. The intent of a heuristic technique is to provide a satisfactory solution with minimum effort.

Several heuristic techniques have been developed to level resources (Harris, 1990; Levy, Thompson, and Wiest, 1962; and Martinez and Ioannou, 1993). The two most difficult tasks in resource leveling are determining (1) the order in which and the amount that the noncritical activities are to be delayed and (2) the final combination of activity times that provides the most reduction of resource variations in the profile. One of the more common heuristic-based techniques used to accomplish these tasks is called the *Burgess method* (Burgess and Killebrew, 1962). This method is a systematic approach to order activities and reduce peak resource demand.

To accomplish the first of the two tasks above, the Burgess method basically uses a heuristic that starts with the last or latest noncritical activity in the project and progresses backward to the earliest activity. Activities are delayed on a period-by-period basis, and then the resource profile is tested for a reduction in resource fluctuation. The technique that is used to test the effectiveness of this reduction (task 2 above) is the *sum of squares*. The sum of squares is a measure of the resource requirements for each time period (*days* in our example) in the project. While the sum of the daily resource requirements is constant for any scheduled project, the sum of the squares of the daily requirements decreases as the peaks and valleys are decreased or leveled. Further, this sum reaches a minimum for a schedule that is level (or as level as can be obtained) for the project in question (Moder, Phillips, and Davis, 1983). The numeric value of the sum of squares is given by

$$Z = \sum_{i=1}^{T} y_i^2 \tag{6.1}$$

where Z = sum of squares for the time period i
T = project duration
y_i = sum of resource requirements of activities performed at time unit i

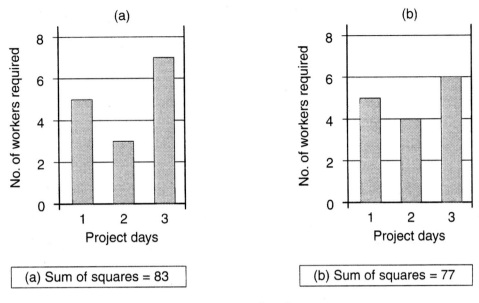

FIGURE 6.7   Sum of squares in resource leveling

A simple example proves the assertion in the sum of squares method. The resource profile in Figure 6.7 shows the number of workers required on the first 3 days of a project. In Figure 6.7(a), the original early start schedule is given with no resource leveling and an associated sum of squares (Z) value of 83. A shift in noncritical activities results in the resource profile in Figure 6.7(b) with an associated Z value of 77. The leveling process lowered the sum of squares by a magnitude of 6. It's clear that the resource profile in Figure 6.7(b) is more level than the profile in Figure 6.7(a).

Burgess resource leveling is relatively simple to apply manually and is appropriate for a variety of project conditions and resource constraints. The systematic procedure is explained in the following steps.

1. Develop the AON network for a time-constrained schedule.
2. Compute the early and late start and finish times and the float for each activity.
3. Compute the sum of squares for the time-constrained, early start schedule.
4. For the noncritical activities only, list the activities in reverse sequential order according to their late start times. This step develops the *reverse late start order* where the noncritical activity with the latest LS is first in the order and so on.

5. Start with the first activity in the reverse late start order (i.e., the activity with latest late start).

6. For the current activity, schedule (i.e., delay) the activity period by period and compute the sum of squares for each period. Schedule the activity in the time period with the minimum sum of squares of resource requirements. If the sum of squares is tied for any two (or more) time periods, schedule the activity as late as possible to provide all available float to all preceding activities. If the activity has no noncritical predecessors, schedule the activity in the earlier period of the tie. In this situation, the float should not be wasted without a reduction of the sum of squares.

7. Holding the previous activity fixed in the schedule (from step 6), go to the next activity in the reverse LS order and repeat step 6. Continue selecting and scheduling noncritical activities in this manner until all activities in the reverse late start order have been analyzed.

Let's apply the Burgess resource leveling procedure to the example project shown in Figures 6.2, 6.3, and 6.5. The network is developed in Figure 6.2. Forward and backward pass computations of the network result in the start, finish, and float times for each activity, as given in Table 6.1. The sum of squares for the time-constrained, early start schedule is computed using Equation 6.1, where the daily resource requirements for all activities are totaled, those totals squared, and the squares summed for a final sum of squares, Z. For this schedule, Z is equal to 852 (Figure. 6.8).

The noncritical activities in the example project are B, D, E, G, and I. The reverse late start order of the noncritical activities (from Table 6.1) is I (LS = 17), E (13), G (9), B (8), and D (5). Activity I is first for shifting, with 1 day of float available. Delaying activity I by 1 day does not change the sum of

TABLE 6.1 Example Project Start, Finish, and Float Times

ACTIVITIES	EARLY START	EARLY FINISH	LATE START	LATE FINISH	FLOAT
A	0	2	0	2	0
B	2	7	8	13	6
C	2	8	2	8	0
D	2	6	5	9	3
E	7	10	13	16	6
F	8	16	8	16	0
G	6	14	9	17	3
H	16	19	16	19	0
I	16	18	17	19	1
J	19	22	19	22	0
Project duration = 22					

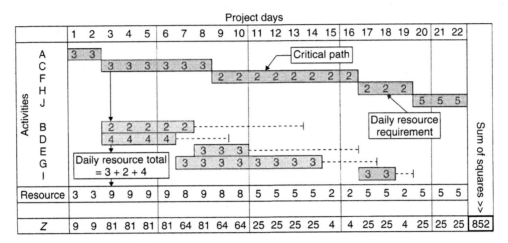

FIGURE 6.8　Sum of squares computations—original schedule

squares, $Z = 852$. But I is preceded by another noncritical activity; thus I is delayed by the 1 day in order to free up more float time for activity G. The logic for delaying I in favor of G is clarified by examining the early and late start and finish times of the two activities. If I is not delayed and kept at its early start (ES) of day 16, then the latest G can finish is day 16 since G precedes I. This results in an available float for G of 2 days. If I is delayed to its LS, then G can finish on its LS (i.e., day 17). Consequently, G gets all the available float of 3 days.

The next activity in the reverse LS order (i.e., activity E) is selected, and step 6 in the Burgess procedure is repeated. The process of selecting activities to delay in reverse LS order continues until all noncritical activities have been analyzed and a Z value calculated for each delay of every activity. The results of this process are given in Table 6.2.

Once all actions have been completed in Table 6.2, a final resource-constrained schedule and associated resource profile are developed, as shown in Figure 6.9. A comparison of the schedules before (Figure 6.5) and after (Figure 6.9) resource leveling is given in the side-by-side resource profiles in Figure 6.10. Some of the results of the resource leveling procedure that can be observed in Figure 6.9 are noteworthy:

1. All available float time has been used on activities E and I, making the activities critical.
2. Available float time has been reduced on all activities except D.
3. Activities B and D are no longer in parallel (i.e., occur simultaneously).
4. The resource-constrained schedule is significantly different from the time-constrained schedule, even for this relatively small project.

TABLE 6.2 Burgess Sum of Squares

ACTIVITY	DELAY ACTION	Z	MINIMUM Z	CONCLUSION
–	No delay, early start schedule	852	—	—
I	Delay I by 1 day	852	$852 = 852$	Delay I 1 day
E	Delay E by 1 day	846		
	Delay E by 2 days	846		
	Delay E by 3 days	846		
	Delay E by 4 days	846		
	Delay E by 5 days	828		
	Delay E by 6 days	810	$810 < 852$	Delay E 6 days
G	Delay G by 1 day	810	$810 = 810$	Delay G 1 day
	Delay G by 2 days	822		
	Delay G by 3 days	822		
B	Delay B by 1 day	806		
	Delay B by 2 days	798		
	Delay B by 3 days	790		
	Delay B by 4 days	782	$782 < 810$	Delay B 4 days
	Delay B by 5 days	790		
	Delay B by 6 days	786		
D	Delay D by 1 day	798	$798 < 782$	Do not delay D

6.5   RESOURCE ALLOCATION

In the previous section, resource leveling was applied to smooth or level out the resource profile. Often, the project planner is faced with the problem of limits on resources, and leveling does not sufficiently reduce the peak demand for particular resources. For the "leveled" profile in Figure 6.9, eight units of the resource (e.g., number of skilled workers) are required for 3 days of the 22-day project. If only seven skilled workers were available for the project, a conflict would exist that could not be resolved by a resource leveling method. In this case, alternative methods must be employed to schedule the project to meet these resource constraints.

Resource allocation is a method where maximum limits are set for each resource and specific amounts of resources are allocated to project activities according to certain scheduling heuristics. The intent is to schedule the activities so that resource limitations are not exceeded while keeping the project duration at a minimum. In other words, the project finishes in as short a time as possible under the restrictions of activity precedence, resources limits, and time constraints, in that order. As a result, the project duration or completion date may be extended to keep resources within specified limits.

A scheduling heuristic determines the activities to be executed first and those to be postponed if the total resource requirements of two or more simultaneous

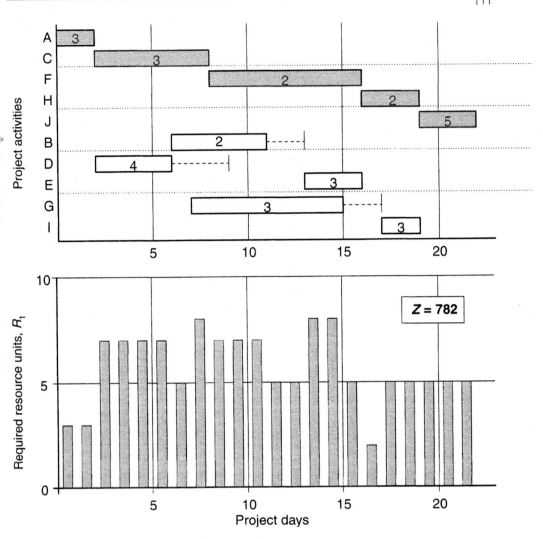

FIGURE 6.9   Resource profile after leveling

activities for a given time period exceed a maximum limit. As stated at the end of Section 6.2 in this chapter, in order to apply the resource management procedures described here, the assumption is made that no single project activity will exceed the maximum limit available for any resource.

   A number of heuristic-based procedures for solving resource allocation problems are available, including the Brooks, MS^2, and SPAR-1 methods (Hinze, 1998, pp. 144–151; Moder et al., 1983, pp. 208–217; and Wiest and Levy, 1977, pp. 112–123). Basically, there are two general solution approaches for resource allocation problems—series and parallel. A *series* approach uses a heuristic to rank or sort all project activities in a single group, and then the activities are

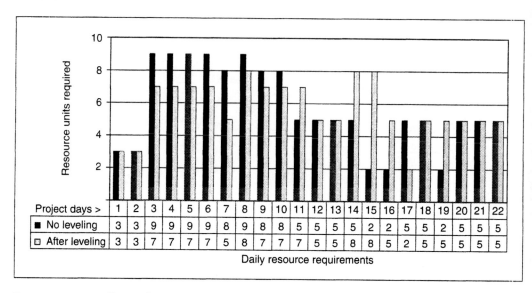

FIGURE 6.10 Effect of Burgess leveling on resource profile

It is important to note that there are primary differences in the resource leveling and resource allocation stages described here. Resource leveling involves the efficient use of noncritical activity float time to redistribute conflicting resources while not altering the time-constrained project duration (i.e., *the project completion date is fixed*). Resource allocation involves the allocation of resources that are limited by scheduling activities according to heuristics. Resource allocation scheduling often results in an extended project duration—one that is longer than the time-constrained project duration (i.e., *the project completion date is not fixed*).

scheduled one at a time (i.e., in series). An attempt is made to start activities in rank and precedence order at the early start times, but activities can be delayed if sufficient resources are not available. The series approach depends on the assumption that activities cannot be split (i.e., work cannot be interrupted during normal work hours) once they are started.

A *parallel* approach ranks all activities in a given time period in order of priority, and resources are allocated according to this priority. When resource limitations (due to concurrent activity requirements) prohibit an activity being scheduled, it is delayed until the next time period. At each subsequent time period, the activities potentially beginning on that period are rank-ordered and the process continued. There are two distinct differences between the series and parallel methods: (1) Activities are rank-ordered only once in the series method and multiple times in the parallel method, and (2) activities, once started, cannot be interrupted in the series method (i.e., no activity splitting) and can be interrupted in the parallel (i.e., activity splitting allowed).

Both the series and parallel methods employ heuristics to schedule activities. The heuristics are based on factors such as the *earliest late start, minimum total float, criticality in the original time-constrained schedule, largest number of resources,* or *activity has already started* (top priority in the series method). In both methods, activity precedence must be observed. Activities are typically brought into the schedule at their early start times as soon as their predecessor(s) is finished, if sufficient resources are available. If sufficient resources are not available for two competing activities in one time period, one or more of the heuristics listed here are applied to "break the tie" between the two activities.

The parallel approach is somewhat more complex and requires more computational time to rank-order activities at each time period after placement of activities in the previous time period. However, the parallel approach seems to be the more widely used method for commercial software that performs resource management because of some of the limitations of the series method (e.g., activity splitting). For the manual solution given here, the series approach will be used because it's easy to understand and yet fully demonstrates the conceptual framework of the allocation process. The series method for resource allocation is performed with the following steps:

1. Develop the AON network for a time-constrained schedule.
2. Compute the early and late start and finish times and the float for each activity.
3. Develop a series-method, resource allocation table and list all project activities vertically.
4. Develop a list of activities in sequential order according to their late start times, where the earliest LS is first in the order and so on. This step develops the *late start order*.
5. Start with the first time period and step progressively through the duration of the project period by period to the project finish.
6. Schedule activities to start at their early start time as soon as their predecessors have been completed. Once an activity has started, do not interrupt it until finished. Give it the highest priority (i.e., no activity splitting).
7. If two or more activities in one time period exceed the maximum limit for any resource, schedule the activity with the earliest late start.
8. If two or more activities are tied for earliest late start times, give priority to the activity with the *least total float*, then to the activity with the *largest number of resources*, then to the *original time-constrained critical path*, and finally to *alphanumeric order*.
9. Schedule activities period by period and compute the total number of resource units used within each period without exceeding the maximum limit for any resource.

The example project from the resource leveling problem in the last section (refer to Figures 6.2 and 6.5) is used for this resource allocation problem. Assume that the maximum resource requirements cannot exceed 7 units per day (e.g., skilled workers). The late start times (from Table 6.1) for each activity are A-0, B-8,

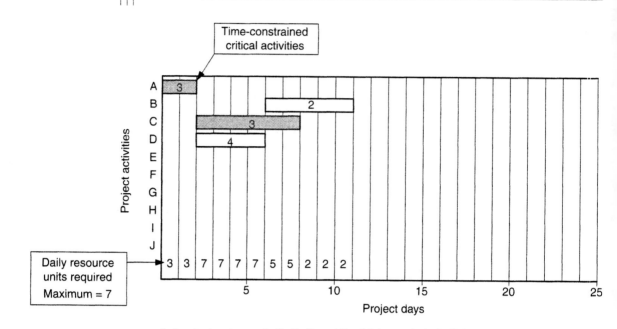

Late start order — A; C; D; B and F, which are tied; G; E; H; I; and J

FIGURE 6.11   Resource allocation series method—scheduling activities A, B, C, and D

C-2, D-5, E-13, F-8, G-9, H-16, I-17, and J-19, yielding a *late start order* of A; C; D; B and F, which are tied; G; E; H; I; and J. On day 1 of the project, activity A is scheduled because it's the only eligible activity by precedence Figure (Figure 6.11). The daily resource requirements for activity A (i.e., 3 skilled workers) are shown at the bottom of Figure 6.11. Once activity A is finished, activities B, C, and D are eligible for scheduling. If all three activities are started, the number of skilled workers required (i.e., 2 + 3 + 4 = 9) exceeds the maximum number of workers available (i.e., 7). Therefore, schedule the activities one at a time according to the late start order. Activity C is scheduled first and then D. This results in the number of skilled workers required for day 3 to be 5, and activity B must be delayed until D is finished at the end of day 6. At the beginning of day 7, activities A and D have finished and activities B and G are eligible to start (refer to Figure 6.2 for precedence). However, both can't be scheduled on day 7 due to exceeding available resources. Thus, B is scheduled before G (Figure 6.11) because B comes before G in the late start sort.

Activity C finishes at the end of day 8, and F and G are eligible to start at the beginning of day 9. Since their concurrence does not exceed the maximum resource availability, they both can start on day 9 (Figure 6.12). While precedence allows activity E to start on day 12 after B is finished, resource requirements exceed availability. Therefore, E must be delayed until at least day 17 (i.e., finish of F and G). On day 17, both E and I are eligible to start (activity H is not eligible because of precedence). Since their concurrence does not exceed the maximum resource

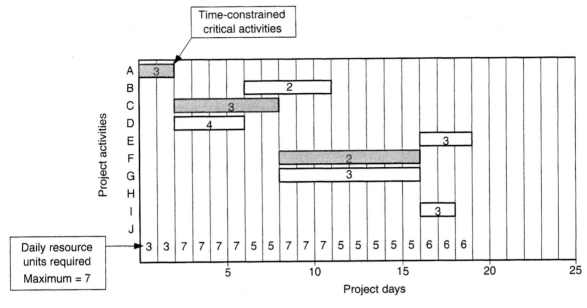

Late start order — A; C; D; B and F, which are tied; G; E; H; I; and J

FIGURE 6.12 Resource allocation series method—scheduling activities E, F, G, and I

availability, both E and I can start on day 17. After activity E finishes on day 19, H can start on day 20. At the finish of H on day 22, the final project activity J can start on day 23 and finish on day 25. As a result of implementing the resource allocation series method, the project duration is extended to 25 days, 3 days more than allotted in the original 22-day time-constrained schedule (Figure 6.13).

6.6 SUMMARY

Resource management is one of the most important aspects of construction project management in today's economy because the construction industry is resource-intensive and the costs of construction resources have steadily risen over the last several decades. Often the project planner utilizes the time-and-precedence–based schedule as a basis for the management of resources for the project. The schedule is developed considering only the constraint of time and precedence among activities (time constraints) and then concentrates on the resource requirements (resource constraints) that are estimated for each project activity, particularly equipment, materials, and labor.

In a resource-constrained schedule, resource units are limited at some maximum level that, for any given resource, is assumed to be less than required during one or more of the time periods within the project. The trends for a resource-constrained

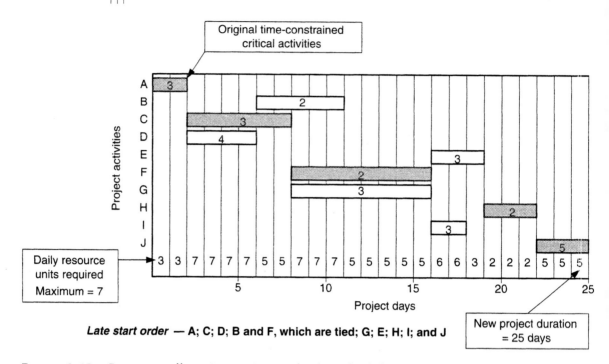

FIGURE 6.13   Resource allocation series method—scheduling activities H and J

schedule are commonly (1) the project activity float time is reduced, consequently making noncritical activities more critical; (2) early and late schedules tend to be no longer unique (for noncritical activities); (3) float is a function of both precedence relationships and resource limitations; and (4) the time-constrained critical path(s) may be different in a resource-constrained schedule.

Resource management techniques (i.e., scheduling, leveling, and allocation) must be employed to distribute resource usage period by period in order to minimize resource variations over time. *Resource scheduling* provides a profile of resource usage during the duration of the project. *Resource leveling*, often applied when the project completion date is specified and fixed, uses the float time of noncritical activities to redistribute activity start and finish dates in order to eliminate or at least reduce resource conflicts. With *resource allocation*, which frequently extends the project duration, guidelines are used to assign limited resources to project activities according to their level of criticality.

CHAPTER 6   QUESTIONS/PROBLEMS

1. *Team Activity* - For the house construction project in Problem 4 in Chapter 5, realistically estimate the number of carpenters and skilled laborers required to complete each job in the project. Subcontract the

installation of the HVAC, plumbing, and electrical systems. Determine the daily resource requirements (two types of resources—carpenters and laborers) for the project and generate a resource profile on a daily time scale.

2. For the resource requirements of the house construction project determined in Problem 1, use Primavera's SureTrak software to define these resources, and enter the requirements for each activity. Generate a resource profile accompanying a bar chart of the project to appropriately fit within two pages (8.5-in. by 11-in. page size) and print both on the two sheets. The bar chart should appear at the top of the pages and the resource profile at the bottom. Cut and paste the chart and profile to show one continuous drawing for each.

3. Generate two resource profiles—one based on an early start schedule and one based on a late start schedule—for the project shown in the figure below. Resource requirements and durations are provided for each activity.

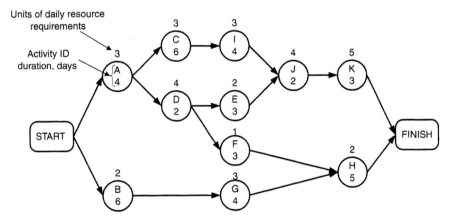

4. Using the early start schedule for Problem 3 above, use the sum of squares method to level resources for the project.

5. For the schedule shown in the figure below, use the sum of squares method to level resources for the project.

DAYS	1	2	3	4	5	6	7	8	9	10	11	12	13	14
Activities A,D	8	8	8	9	0	0	0	0	0	0	0	0	0	0
B,E,G	6	6	6	6	6	6	6	6	4	4	4	4	4	4
F	0	0	0	0	0	9	9	9	9	0	0	0	0	0
H	0	0	0	0	0	0	0	0	0	4	4	0	0	0
C	4	4	4	4	4	4	0	0	0	0	0	0	0	0
RESOURCE	18	18	18	19	10	19	15	15	13	8	8	4	4	4
RESOURCE ^2	324	324	324	361	100	361	225	225	169	64	64	16	16	16

6. For the schedule shown in the figure below, use the serial method to allocate resources such that the maximum daily resource usage is no more than 16 for the project.

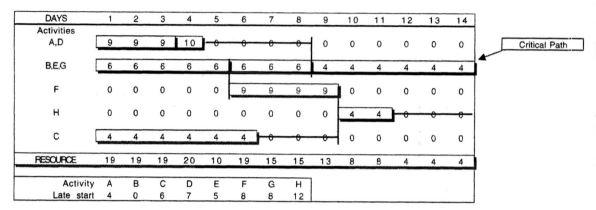

DAYS	1	2	3	4	5	6	7	8	9	10	11	12	13	14	
Activities A,D	9	9	9	10	0	0	0	0	0	0	0	0	0	0	Critical Path
B,E,G	6	6	6	6	6	6	6	6	4	4	4	4	4	4	
F	0	0	0	0	0	9	9	9	9	0	0	0	0	0	
H	0	0	0	0	0	0	0	0	0	4	4	0	0	0	
C	4	4	4	4	4	4	0	0	0	0	0	0	0	0	
RESOURCE	19	19	19	20	10	19	15	15	13	8	8	4	4	4	

Activity	A	B	C	D	E	F	G	H
Late start	4	0	6	7	5	8	8	12

REFERENCES

See References on page 199.

PROJECT TIME REDUCTION/ TIME-COST TRADE-OFFS

O B J E C T I V E S

This chapter provides knowledge in the areas of:

❑ Direct and indirect project costs
❑ Need for reduction in project duration
❑ Time reduction methods
  • Without increased costs
  • By expediting
  • Using heuristics

7.0   OVERVIEW

This chapter describes methods to reduce project duration that minimize project costs. The chapter begins with a basic explanation of direct and indirect project costs including a description of the basic types of schedules relative to the relationship of project duration and total project costs. The chapter discusses reasons for the need to reduce project duration and introduces the time-cost trade-off process. The focus of the chapter then shifts to physical and logical time reduction limitations and discusses methods to reduce project duration without increasing total project costs. Finally, the chapter examines the time reduction method of expediting or buying time along the critical path, demonstrating the method through examples. This section includes an explanation of time-cost relationships for project activity duration reduction as well as cost slope computations and comparisons.

7.1 CONSTRUCTION PROJECT COSTS

In the construction planning process, the final plan/schedule is developed under one primary objective—to optimize both time and money while maintaining quality and safety. In other words, the project should be completed within the least duration, at minimal cost, under safe working conditions, resulting in the highest-quality product. This is a difficult task because least duration and minimal cost tend to conflict with high quality and safety. The task is further complicated by the fact that activity processes within construction projects are somewhat unique. When comparing processes among different projects and among activities within a particular project, a number of alternative resources and sequences can combine to complete a project. Furthermore, each set of alternatives results in a somewhat unique project plan.

Construction planning and scheduling provide a work plan for the activities to be accomplished in order to complete a project. The planning and scheduling process determines start and finish times for each activity and an estimated project duration, based on activity time estimates. These estimates are considered a result of "normal," or average, time estimates for activity durations. Normal time estimates and associated resource allocation generally result in an efficient schedule but not necessarily the "least-cost" or "least-time" schedule. The schedule is a compromise between cost and time to accomplish the project tasks. Actually, several types of schedules can be developed depending upon the intent of the planner, as follows:

- ❏ **Normal**   Average time estimates and resource allocations are applied here under "normal" working and job conditions. The normal planning process minimizes time and cost together but doesn't always result in a least-cost or least-time schedule.
- ❏ **Least cost**   In order to minimize the total project cost, both the direct and indirect costs must be minimized. In some instances, increasing the direct cost to shorten one or more activities (i.e., less than the normal duration) may reduce the project duration and thus the indirect cost sufficiently to reduce the total project cost and provide a least-cost schedule.
- ❏ **Least time**   The duration of the project can be reduced further than the least-cost schedule, but at a higher cost. Adding more resources or changing schedule logic, both resulting in higher costs, can shorten the activities along the critical path(s). Collapsing, or "crashing," these activities generates the least-time schedule.

Notice that both the least-cost and least-time schedules involve the shortening of one or more activities resulting in a reduction in project duration. The purpose of this chapter is to consider these schedule changes in order to reduce project duration—to examine the relationship between project duration and total project costs.

Total project costs can be lumped into two primary categories—direct costs and indirect costs. Direct costs are those costs accrued from the use of resources to accomplish project activities. They are a result of quantities and prices of materials, time spent by workers, and duration of use and cost of equipment. The direct cost for any given activity varies depending on the type and amount of resources that are applied. For example, the assumption of a normal-size work crew to complete an activity results in a certain duration and associated direct cost of labor. Assigning more workers to this activity normally results in a shorter duration, but at an increased cost and possibly reduced quality. The total direct cost of a project is simply the sum of the direct costs for all project activities.

Indirect or overhead costs are costs resulting from the support of field work (job overhead) and the support of the contracting company (company overhead). Neither is associated with particular project activities. This support includes cost items such as financing; insurance; taxes; office personnel, services, and supplies; inspection; testing; and resources not included in construction. Indirect costs are typically incurred regardless of job productivity and tend to increase as project duration increases. Thus, any increase in project duration directly affects profitability. The magnitude of indirect costs makes consideration of the relationship between project duration and total project cost essential. Indirect costs are added to the direct costs for a total project cost.

7.2 NEED FOR REDUCTION IN PROJECT DURATION

Chapter 6 contains a thorough discussion of time- and resource-constrained construction schedules. This discussion concentrates on time constraints, but the relationship between time and resource allocation becomes more evident here. For example, adding more workers to a "normal" work crew may reduce the "normal" estimated time for a particular activity. Assuming the activity to which more workers are added is critical (i.e., occurs along the critical path), then a reduction in this activity duration results in a reduction in the project duration by the same amount. Time constraints in a construction project arise for a number of reasons. Some of the more common are:

- ❏ The contractor makes an unrealistic, elevated prediction of one or more activity durations, thereby falsely elevating the project duration in the schedule.
- ❏ The owner contractually requires a completion date for the project which is earlier than the scheduled finish date.

❑ Weather delays and/or delays in the delivery of materials cause the project to get behind schedule.

❑ An early-completion bonus or late-completion penalty is written into the contract.

❑ A least-cost schedule is sought in order to minimize the total project costs.

The purpose of least-cost scheduling is to balance direct and indirect costs. Indirect costs tend to decrease as project duration decreases, but this may be at the expense of increased direct costs of a faster but less efficient set of resources and sequence of operations. There are other negative effects of actions taken to reduce project duration. For example, increasing the number of workers (beyond the normal work crew) on an activity or set of activities to reduce duration may cause crowded work conditions and lower worker productivity. Work quality and safety may suffer.

A preference may be given to a schedule that has higher estimated direct costs if it reduces project duration and related indirect costs such that the total project cost is decreased. There can, therefore, be a trade-off between alternative schedules. Evaluating alternatives in this fashion is called a *time-cost trade-off* comparison. In this chapter, methods to reduce project duration in the most cost-efficient manner are examined.

7.3   TIME REDUCTION WITHOUT INCREASED COSTS

The time reduction of activities and project duration is constrained by the amount of time available for reduction. These constraints or reduction limits are based on the work to be accomplished, activity precedence relationships, and relationship of the critical path to other noncritical paths. The constraints are summarized as follows:

❑ **Physical limit**   This is the absolute maximum reduction of a critical activity time to get that activity to its minimum duration needed to accomplish the activity work, regardless of additional resources. For example, an activity involving the placement and curing of concrete has an absolute minimum time due to the time needed for concrete hardening.

❑ **Logical limit**   This is the absolute maximum reduction of a critical activity time without making critical (i.e., using up all available total float) an adjacent noncritical path. Any activity along the critical path whose duration is shortened affects other parallel noncritical paths. For each reduction in time period along the critical path, the total float of all noncritical parallel paths is reduced the same amount. Eventually, if the critical path is reduced sufficiently, the total float of another noncritical path will be used up and that parallel path will become critical. To obtain a further reduction in project duration, activities along both critical paths would then need shortening.

Reducing project duration or time without increasing total project costs can be accomplished generally by (1) developing the least-cost schedule and/or (2) carefully analyzing and modifying the "normal" project schedule. The intent in both these processes is to reduce project duration while maintaining or lowering overall project costs.

7.3.1 LEAST-COST SCHEDULING

As stated earlier, the least-cost schedule minimizes the total project costs by trading an increase in direct costs to shorten one or more critical project activities, resulting in a reduced project duration and an associated decrease in indirect costs. Indirect project costs typically increase linearly as project time increases (Figure 7.1).

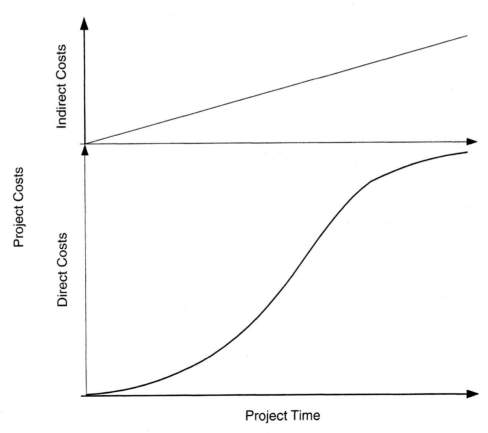

FIGURE 7.1   Project costs as a function of time

A reduction in project duration tends to reduce the indirect project costs. Therefore, any decrease in project duration tends to decrease indirect costs linearly. If the savings in indirect costs resulting from a decreased project duration are more than the direct costs to shorten one or more critical activities, then the total project costs are reduced, resulting in a least-cost schedule.

Let's look at an example of least-cost scheduling from the road bridge foundation schedule shown in Figure 7.2. In this example the indirect cost for the road bridge project is $350 per day. The project duration can be shortened by 1 day by reducing any critical activity by that same time. The critical activity *Pour footing #1* can be expedited (i.e., its duration shortened) from 3 to 2 days by adding another laborer to the activity. The direct cost for the laborer is as follows:

$$2 \text{ days} \times 8 \text{ hours per day} \times \$20/\text{hour (salary and benefits)} = \$320$$

By adding the laborer for the direct cost of $320, the project duration is reduced by 1 day, saving the project $350 in indirect costs. Therefore, the change is justified based on developing a least-cost schedule. This example does not take into consideration whether the laborer is available nor how the added laborer may encumber space and/or affect work quality and safety.

7.3.2   Analysis/modification of "normal" project schedule

Before attempts are made to reduce the project duration using methods that increase the direct costs of the project, a careful and thorough analysis of

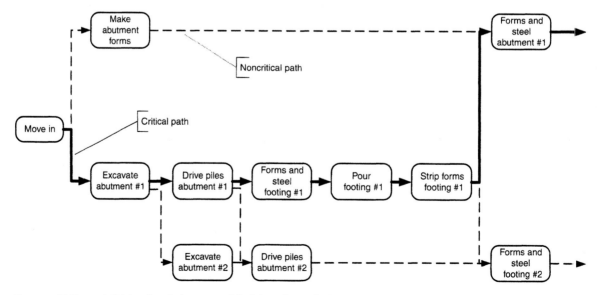

Figure 7.2a   AON schedule—road bridge foundation

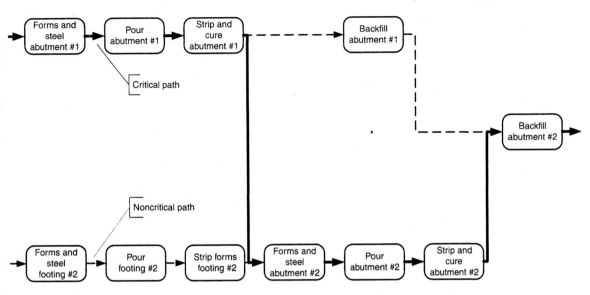

FIGURE 7.2b AON schedule—road bridge foundation

duration reduction in the existing schedule should be considered. When developing the original schedule, the project planner must rely on his or her personal experience and "best judgment"—and/or the knowledge and experience of others in the company—for time and cost estimates. These estimates, while based on realistic assumptions by all involved, are completed under circumstances that may have been somewhat uncertain. It is recommended that the following actions that do not increase project costs be taken to reduce project duration prior to increasing costs for that purpose:

1. Carefully examine all critical activities to assure their durations are achievable and accurate. Most importantly, are any duration estimates *overestimated?*

2. Study job logic. Is it possible to *rearrange critical activities* to achieve a schedule with a shorter duration?

3. Carefully check the serial nature of critical activities. If possible, run critical activities in parallel or overlap portions of critical activities. Must a particular activity be *completely* finished before the next activity can be started?

4. Consider *subcontracting* one or more critical activities to run in parallel. Are two critical activities in series because of limited resources rather than the physical order of the work? In some cases, the cost of a subcontractor may be competitive with internal costs.

7.4   TIME REDUCTION BY EXPEDITING (BY INCREASING COSTS)

As stated earlier, the main purpose of this chapter is to examine methods to reduce project duration with no or at least minimum increase in total project costs. We've discussed methods to accomplish this without increasing total project costs. Several methods of time reduction by expediting or by minimally increasing total project costs have been reviewed in the literature (Moder, Phillips, and Davis, 1983, pp. 237–253; Hendrickson and Au, 1989, pp. 56–62; and Wiest and Levy, 1977, pp. 62–85). This section examines a method to reduce project duration by *buying time along the critical path*, which minimizes the increases in total project costs. This procedure incrementally reduces the duration of critical activities (and thus project duration) in order of least cost. To understand this incremental reduction of project duration, it is first necessary to examine the relationship of activity duration and direct cost. Three types of relationships are presented graphically (Figure 7.3) to describe the interaction of time and cost as follows:

❑ **Continuous linear [Figure 7.3(a)]**   This is a simplistic explanation of the time-cost relationship where a period-by-period reduction in duration for any given activity costs exactly the same from one period to the next. The added direct cost for each time period reduction is consistent. While somewhat unrealistic, this relationship can be applied within a few limited increments of duration for certain activities.

❑ **Continuous stepwise linear [Figure 7.3(b)]**   For the period-by-period reduction of most project activities, this relationship is often most realistic. In this relationship, as more and more reductions are expedited, the cost of each subsequent reduction increases above that of the previous period. The cost slope increases with each time period reduction, indicating progressively larger direct costs. This relationship is sometimes represented by a "convex" curve rather than stepwise linear; however, the stepwise linear is actually more realistic since the costs change on a period-by-period basis.

❑ **Noncontinuous [Figure 7.3(c)]**   In a few instances, it is possible to reduce the duration of certain activities for only one block of time rather than on a period-by-period basis as with the continuous linear and continuous stepwise linear relationships. An example is hiring a subcontractor to complete a segment of the project. The subcontractor's expertise allows the duration of one or more project activities to be reduced by a certain block of time [e.g., 3 days in Figure 7.3(c), from 11 to 8 days]. But it is not likely that the subcontractor would negotiate only a portion of the work such as on a daily basis. Another example would be paying a premium for early delivery of material or equipment to reduce a critical activity by a certain block of time, with no other possibility.

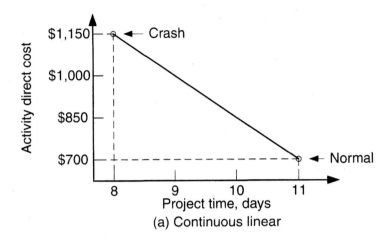

(a) Continuous linear

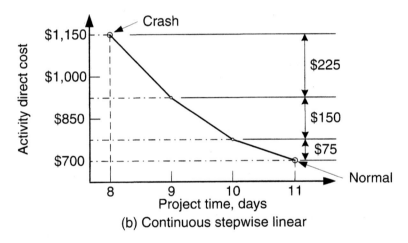

(b) Continuous stepwise linear

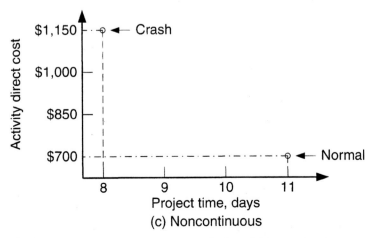

(c) Noncontinuous

FIGURE 7.3 Activity time-cost relationships

In the procedure to *buy time along the critical path*, project duration is reduced incrementally along the critical path within the *physical* and *logical limits* of the critical activities and path as detailed above. In other words, activities along the critical path are examined for potential reduction, or "crashing," and those activities that can be expedited (shortened) are identified. Activities are selected for crashing, and the direct costs of crashing are determined. The activities are order-ranked for reduction by least direct cost.

This procedure effectively reduces project time by typically providing more resources to those critical activities that can be obtained at the least cost. Determination of the least-cost activities requires careful examination of the critical and nearly critical paths (i.e., those parallel noncritical paths with minimal total float time). The key element of this procedure is identifying activities where time can be bought at least cost. A time-cost comparison factor, the activity *cost slope*, can be used for this purpose. It is defined in Equation 7.1 and graphically illustrated in Figure 7.3(a).

$$\text{Cost slope}_i = \left[\frac{\text{normal cost}_i - \text{crash cost}_i}{\text{normal duration}_i - \text{crash duration}_i} \right]$$

where
cost slope_i = cost per unit time reduction for activity i
normal cost_i = cost of activity i occurring over the estimated normal duration for activity i
crash cost_i = cost of reducing activity i for the crash duration

Buying time along the critical path first requires the development of a normal schedule in which all activities start at the early times with normal resources and associated costs. The normal schedule is expedited (i.e., duration reduced) by reducing the time of one or more activities along the critical path at additional direct cost. If this added direct cost is less than the savings in indirect costs that result from a reduction in the project duration, then a least-cost schedule results. The procedure continues in a stepwise manner such that total project costs are reduced or at least held to a minimal increase. The intent is to identify those activities with the *minimum cost slope* so that those activities with the least direct cost per time period are selected.

The main problem in this procedure is determining which activities to investigate for potential reduction and how far to take the schedule-shortening process. The question of how much shortening is appropriate depends on the necessity for the reduction. The need for reduction is dependent on the time constraints placed on the project as detailed in Section 7.2 above. The investigation of activities for potential reduction depends on the minimum cost slope. For projects with few activities, all critical activities can be examined (i.e., a cost slope determined). For example, Figure 7.4 illustrates a relatively simple, small project, and Table 7.1 gives information for expediting the critical activities for that project.

The order of expediting the critical activities can be obtained from the cost slope data in Table 7.1. The minimum to maximum cost slope order is E, C, A,

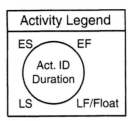

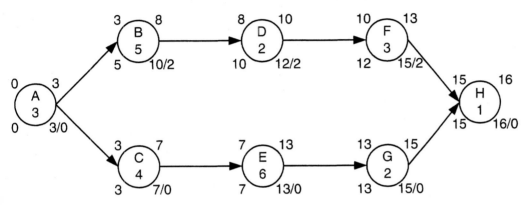

FIGURE 7.4 AON schedule for expediting example

TABLE 7.1 Expediting Information for Example Project

ACTIVITY	NORMAL DURATION	CRASH DURATION	NORMAL COST	CRASH COST	PHYSICAL LIMIT	LOGICAL LIMIT	COST SLOPE (COST PER DAY)
A	3	2	$700	$1,000	1	3	$300
C	4	2	$1,200	$1,700	2	2	$250
E	6	4	$500	$800	3	2	$150
G	2	1	$600	$1,000	1	2	$400
H	1	1	$300	$300	0	0	—
B							
D			Noncritical activities not considered for expediting.				
F							

and then G. It can be observed that a maximum reduction in either activity C or activity E of 2 days would reach the logical limit of both activities. Activity E is the first reduction in the cost slope order, and if E is reduced by 2 days to its crash duration, the noncritical path (i.e., activities B, D, and F) becomes critical. Any further reduction in the project after the 2-day reduction in E requires examination of activities B, D, and F (previously noncritical). The cost slope for B, D, and F is required, and then a reduction of two activities from the adjacent critical paths is necessary to reduce the project duration further. The process is repeated until no further time reduction of critical activities is possible or until the costs of further reduction exceed the savings that result from reducing the project duration.

In this simple example with few activities, a manual examination of all critical activities in the original schedule was feasible. We looked at all possible alternatives by determining the crash cost and duration of each critical activity and then calculated their cost slopes. It was then relatively straightforward to choose critical activities for reduction in a stepwise manner according to minimum cost slope. As the size of projects increases with an accompanying increase in the number of activities and network paths, the procedure becomes more difficult due to the rapid increase in the number of alternatives to evaluate in each step.

For example, the critical path for the road bridge foundation project in Figure 7.2 has 22 activities, 14 of which are critical. While the "best" crashing solution of this project (still relatively simple with only 22 activities) would be to determine the crash duration and cost for each critical activity, the time required to determine the crash duration and cost for each activity may be significant. The project planner likely has many other responsibilities, and so it may not be feasible for him or her to dedicate that much time for a full examination. Clearly, a manual solution of all alternatives is not feasible for large projects. Therefore, other means of choosing critical activities for "crashing" analysis is warranted.

For resource leveling and allocation (Chapter 6), heuristic-based approaches were employed to provide optimal solutions with reasonably minimal efforts. The use of heuristics for the selection of critical activities for crashing is more subjective in this procedure than those in resource management. This is due to the number of mutually supporting and company-dependent factors that influence the decision to crash any given activity. In a previous example in this chapter, the activity *Pour footer #1* (from Figure 7.2) was reduced in time from 3 to 2 days by adding one more laborer. It was noted in the example that the availability of the laborer and space encumbering were not considered, nor was the effect on work quality and safety. These factors must be considered when choosing activities to determine crash duration and cost. The following heuristics may be applied in this choice in the context of supporting and company-dependent factors that influence the choice:

❑ Give preference to the activity where sufficient company or subcontractor resources are available to carry out the crash.

❑ Give preference to the activity that can change the logic of the critical path resulting in a large reduction in project duration.

❑ Give preference to the activity that has the largest potential reduction.

❑ Give preference to the activity with the least effort to determine crash duration and cost.

TABLE 7.2 Expediting the Road Bridge Foundation Project (Figure 7.2)

	EXPEDITING ACTION	TIME REDUCTION	CRASH COST	COST SLOPE (COST PER DAY)
1	Expedite activity *Pour footer #1* by 1 day by adding one laborer.	1 day	$320	$320
2	Fabricate an additional set of abutment forms and start activity *Forms and steel abutment #2* directly after *Strip forms footer #2*.	5 days (change in logic of critical path)	$2,500	$500
3	Expedite activity *Pour abutment #1* by adding more equipment and laborers and working overtime.	1 day	$400	$400
4	Expedite *Pour abutment #2* same as was done with abutment #1 (action 3).	1 day	$400	$400

❏ Give preference to the activity late in the project so as to have less effect on total project schedule.

❏ Choose an activity at random.

Let's look at a crashing example using the road bridge foundation project presented in Figure 7.2. Several expediting actions are examined assuming they satisfy the selection heuristics given above. The information on these actions is provided in Table 7.2.

The original normal schedule for the road bridge foundation resulted in a project of 45 days. From the information in Table 7.2, the expediting actions result in the following stepwise project crash duration and costs. The order of these actions is based on least-cost slope.

❏ **Expediting action 1**   Project duration reduced from 45 to 44 days at a cost of $320.

❏ **Expediting actions 3 and 4**   Project duration reduced from 44 to 42 days at a cost of $800 ($400 per day). Total reduction is now 3 days at a cost of $1,120 ($320 + $800).

❏ **Expediting action 2**   Project duration reduced from 42 to 37 days at a cost of $2,500. Total reduction is now 8 days at a cost of $3,620 ($1,120 + $2,500).

7.5   SUMMARY

A construction project schedule is a compromise between time and cost. Most schedules developed under normal project conditions are typically not either a least-cost or a least-time schedule. Total project costs can be lumped into two primary categories—direct costs and indirect costs. Direct costs are those costs accrued from the use of resources to accomplish project activities. The total direct cost of a project is simply the sum of the direct costs for all project activities. Indirect costs are those resulting from the support of field work (job overhead) and the support of the contracting company (company overhead). Indirect costs are typically incurred regardless of job productivity and tend to increase as project duration increases. Indirect costs are added to the direct costs for a total project cost.

Time constraints in a construction project make project time reduction a necessity in some cases. Common time constraints include unrealistic predictions of activity durations, contractual requirements for early project completion dates, delays caused by weather and/or delivery of materials, early-completion bonuses or late-completion penalties, and the goal of seeking a least-cost schedule to minimize the total project costs.

Reducing project duration without increasing total project costs can be accomplished by developing a least-cost schedule and/or carefully analyzing and modifying the "normal" project schedule. The least-cost schedule minimizes the total project costs by trading an increase in direct costs to shorten one or more critical project activities, resulting in a reduced project duration and an associated decrease in indirect costs. A careful and thorough analysis of the normal project schedule may yield changes in activities or schedule logic to reduce project duration without increasing costs.

Time reduction by expediting is a method to reduce project duration by buying time along the critical path, which minimizes the increases in total project costs. This procedure incrementally reduces the duration of critical activities (and thus project duration) in order of least cost. A time-cost comparison factor, the activity *cost slope*, can be used for this purpose.

CHAPTER 7   QUESTIONS/PROBLEMS

1. Team Activity –Select two critical activities from the three-unit townhouse in Appendix A and determine the realistic direct costs of crashing these activities on a daily basis to their physical and/or logical limits to reduce project duration. Determine the cost slope of each crashed activity.

2. Compare and contrast the *physical limit* and *logical limit* for a project activity.

3. Compare the *continuous linear* and *continuous stepwise linear* time-cost relationships in Figure 7.3 (a) and (b). Explain why the continuous stepwise linear relationship most accurately represents the actual cost of buying time to crash a construction project.

4. Expediting data for the critical activities of a project are given in the table on the next page. Determine the cost slope of the activities, determine the maximum shortening, and rank-order the activities for reduction.

Expediting Information

CRITICAL ACTIVITY	NORMAL DURATION	CRASH DURATION	NORMAL COST	CRASH COST	PHYSICAL LIMIT	LOGICAL LIMIT	COST SLOPE (COST PER DAY)
A	4	2	$800	$1,000	2	3	
C	5	4	$1,300	$1,700	2	2	
D	6	4	$400	$800	2	1	
F	8	6	$700	$1,000	3	1	
G	3	2	$500	$300	1	1	

REFERENCES

See References on page 199.

PROJECT SCHEDULING WITH UNCERTAIN DURATIONS

This chapter provides knowledge in the areas of:

- ❑ Modeling the construction process
- ❑ Uncertainty in activity durations
- ❑ Program evaluation and review technique (PERT) formulation
  - Central tendency
  - Mean and measure of dispersion
- ❑ PERT procedure

8.0 OVERVIEW

This chapter addresses uncertainty in activity durations and describes a method—the program evaluation and review technique (PERT)—to manage this uncertainty. The chapter begins with a discussion of the construction process and introduces the idea of representing this process with a model or schedule of the actual activities to occur. In discussing the model, the chapter examines a limitation of the network model: Whereas the activity duration estimates are fixed and known in the CPM and PN methods, the PERT procedure is a probability modeling method for scheduling projects that have highly variable and uncertain activity durations. Next, the chapter presents a conceptual framework of the PERT process and explains the procedure formulation. Several examples are provided to illustrate the actual procedure.

8.1 MODELING THE CONSTRUCTION PROCESS

Planning and scheduling a construction project presents the contractor with a difficult and challenging problem. The contractor commences the project with a facility design (i.e., technical drawings and specifications), which provides a very detailed "picture" of the end product or completed facility. However, the facility design provides few details about how the contractor must perform the construction activities to complete the work. The contractor must use the technical descriptions of the completed facility to form a plan that executes all aspects of the work to physically construct the facility. This network plan must contain all required resources (i.e., time, materials, equipment, and labor) and must be such that the facility is completed on time and within the estimated budget.

The network model is the tangible result of the network plan used for management purposes. The network model is a schematic representation of the construction process that describes the process of work to accomplish the project. The network model, applied through project scheduling, is only a representation of the actual construction process, and its performance is only as good as the data used for its generation. Modeling a real system is an extensive process in which a problem must be well defined, abstractly represented, and tested. A model developed with inaccurate data not only misrepresents the process, but also may compromise the construction plan.

The purpose of this chapter is to investigate the PERT project scheduling technique that, unlike CPM and PN procedures, attempts to more accurately model the realistic behavior of the duration of project activities. PERT is a probability modeling method for scheduling projects that have highly variable activity durations. In CPM and PN modeling, time estimates for activity durations are fixed and known although the time actually needed to complete any activity is neither. Uncertainty in activity duration estimates is undeniable, particularly during the planning stages of a project. The program evaluation and review technique (i.e., PERT) is the most common formal method that considers time estimate uncertainty and is mostly applied in projects where considerable time variations are expected for activity durations. Activities in research and development projects usually have large time variations, but this is a lesser problem in construction projects. Therefore, PERT is seldom applied in the construction industry. However, the PERT concept is one that deserves investigation. This is particularly true for understanding the varying nature of activity duration estimates and how they are applied in nearly all project scheduling computer software.

8.2 UNCERTAINTY IN ACTIVITY DURATIONS

The project planner determines the length of each activity duration based on the estimated materials to be used and crew production to complete the activity. Historical data from previous projects completed by the contractor may also be a source of data. Each activity duration is a single time estimate (i.e., deterministic) and is fixed for the scheduling of the project. Individual activity durations are adjusted when evidence is presented justifying a change, but ultimately the network is scheduled with fixed activity durations. The forward and backward pass computations attest to the fixed time estimate for each activity, and project scheduling software requires fixed time estimates to analytically determine the critical path(s) and duration of a project.

For any particular activity, a fixed duration value is chosen when, in reality, the duration varies infinitely along a continuum within a minimum and maximum range (Figure 8.1). This fixed duration can be assumed to be the "most likely, "average," or "normal" time that is required for the activity to accomplish the work under normal conditions. Time and/or resource constraints, however, increase the uncertainty of time estimates. These constraints may include:

❑ Adverse or unexpected weather conditions
❑ Low-productivity or maintenance problems with the equipment
❑ High degree of uncertainty in unique construction methods

Activity durations involve estimating time. Time as a random variable is continuous and varies from some minimum to some maximum time. While there is a relatively good probability that the actual time required for most construction activities will be near the average estimate, duration time is not absolutely fixed or known with certainty. Time, as a continuous random variable, is stochastic (i.e., probability-based) and difficult to apply to traditional scheduling computational techniques.

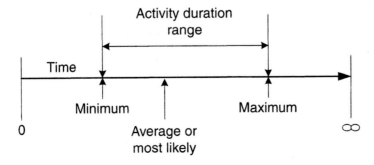

FIGURE 8.1 Activity time as a continuous random variable

8.3   PERT FORMULATION

PERT was developed to aid in producing the U.S. Polaris missile system in record time in 1958. The PERT procedure provides a probability that a project will be completed on or before a specified completion date based on variable time estimates of activity durations. The CPM and PN procedures require known, nonvarying time estimates for each activity in order to analytically determine a solution. PERT, on the other hand, utilizes an average or mean time and a variance (i.e., measure of dispersion or variation) for each activity, calculated from a set of user-provided values. The time estimates used to calculate the mean time and variance in the PERT procedure are assumed to approximate a beta probability distribution (Figure 8.2). In PERT, the formulas for mean time and variance are

$$t_i = \frac{a_i + 4m_i + b_i}{6} \tag{8.1}$$

$$V_i = \left[\frac{b_i - a_i}{3.16}\right]^2 = \frac{1}{10}[b_i - a_i]^2 \tag{8.2}$$

where t_i = average or expected time duration for activity i

a_i = optimistic time estimate for activity i; this estimate is made with the assumption that activity i, if repeated under the same conditions, would result in a duration *less than* the estimate only 1 in 20 times (5 percent)

m_i = most likely time estimate for activity i

b_i = pessimistic time estimate for activity i; this estimate is made with the assumption that activity i, if repeated under the same conditions, would result in a duration *greater than* the estimate only 1 in 20 times (5 percent)

V_i = variance of expected time duration for activity i

The optimistic time is estimated such that there is only a 1 in 20 (5 percent) chance that the actual duration would be less than the estimated time. Similarly, the pessimistic time is chosen with only a 5 percent chance of exceeding the estimate. These 95th percentile limits are used to provide a reasonable estimate from historical data. Thus, there is a 90 percent chance of the actual duration falling between these optimistic and pessimistic estimates. The value of the divisor in the variance equation (Equation 8.2) of 3.2 applies for the 95th percentile limits for a_i and b_i.

The PERT procedure assumes that time estimates for activity durations are independent random variables. While not exactly accurate and one of the problems of the PERT procedure, the assumption does not completely negate

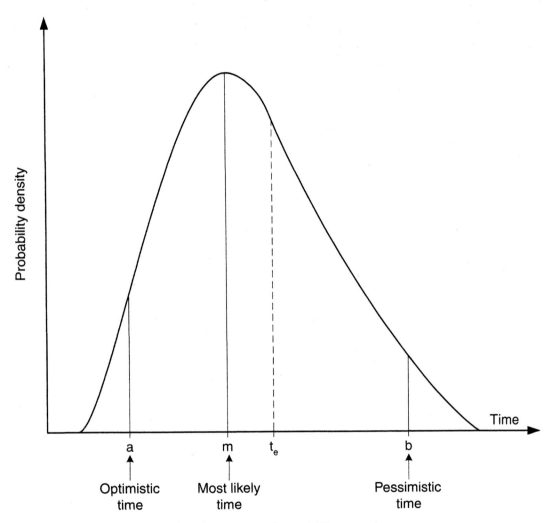

FIGURE 8.2 Typical Beta distribution used in PERT procedure

PERT's validity. Independent random variables tend to follow the central limit theorem, which can be used to describe the behavior of the time estimates, their means, and the associated variations and their effect on project scheduling computations. According to this theorem, the following conventions apply:

1. The mean of the sum of random independent tasks is equal to the sum of their means. When applied to PERT project scheduling, this convention basically states that the expected (i.e., average or mean)

project duration is equal to the sum of the expected durations of the activities along the critical path. This convention is given as

$$\text{Expected project duration } E = t_1 + t_2 + t_3 + \ldots t_n \qquad \textbf{(8.3)}$$

2. The variance of the sum of random independent tasks is equal to the sum of their variances. In PERT, the variance or variation in the duration of the critical path is calculated as the sum of the variances of the activities along the critical path.

$$\text{Expected project variance } V_T = V_1 + V_2 + V_3 + \ldots + V_n \qquad \textbf{(8.4)}$$

3. The shape of the resulting distribution will be "normal," regardless of individual task distributions. In PERT, the resulting mean and variance values of the project duration are normally distributed (i.e., follow a normal distribution) even though the individual time duration estimates for individual activities follow a beta distribution.

8.4   PERT PROCEDURE

For an overview of the PERT procedure, let's consider the following example. The network in Figure 8.3 is provided, with activity time estimate values for the PERT variables a, m, and b given in days [Figure 8.3(a)]. Table 8.1 shows the results of calculations for t_i and V_i by Equations 8.1 and 8.2, respectively, for each activity.

As can be observed in Figure 8.3(b) and Table 8.1, the critical path of the network is A, B, D, and F, with a project duration, E, of 14.3 and project variance, V_T, of 5.27 as per Equations 8.3 and 8.4, respectively, as follows:

$$\text{Expected project duration } E = 2.83 + 4.83 + 3.00 + 3.67 = 14.3 \text{ days}$$

$$\text{Expected project variance } V_T = 0.879 + 2.441 + 0.391 + 1.563 = 5.27 \text{ days}$$

While the results of the PERT computations provide a critical path and project variance, this is little more information than the CPM and PN methods provide. The assumption must be made that the time estimates for each project activity (that is, a, m, and b) are obtained with the knowledge that some amount of uncertainty is present in these estimates. Variations in weather, equipment availability/reliability, personnel and/or material problems, and uncertainty about the work processes may affect activity execution. Therefore, the estimates used to calculate the expected project duration are affected by this uncertainty and any planning that may result from this schedule. This gets to the real intent of the PERT process. PERT is concerned with the problem of estimating the probability of completing a project on time (or within certain time windows) with uncertain time estimates present.

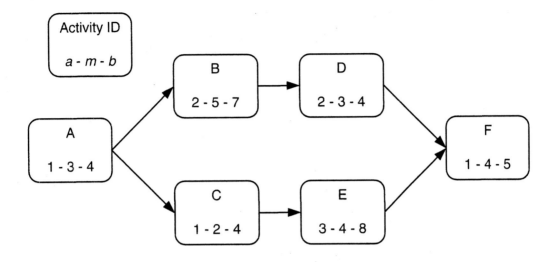

(a) Example network problem

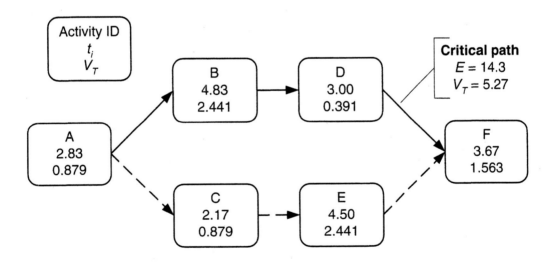

(b) Example network solution

FIGURE 8.3 PERT example network

This highlights the importance in the third convention (stated above) of the central limit theorem. This convention basically states that regardless of the individual activity time distributions, the resulting mean and variance for the project duration will be normally distributed (i.e., follow a normal distribution). Using

TABLE 8.1 PERT t_i and V_i results—example network

ACTIVITY	A	M	B	T_I, DAYS	V_I, DAYS
A	1	3	4	2.83	0.879
B	2	5	7	4.83	2.441
C	1	2	4	2.17	0.879
D	2	3	4	3.00	0.391
E	3	4	8	4.50	2.441
F	1	4	5	3.67	1.563

this convention, the cumulative normal distribution function can be used to predict the probability of the project completion times. The results for this example project are an expected project duration or *mean of 14.3 days* and a project duration *variance of 5.27 days*. The cumulative normal distribution table is given in Figure 8.4. The normal distribution function can be applied through the following two equations within the context of the PERT technique:

$$\Pr \{Z \le z = [T_s - E] \div \sqrt{V_T} \tag{8.5}$$

$$T_s = \left[\sqrt{V_T} \times z \right] + E \tag{8.6}$$

where Z = standard normal random variable
z = normally distributed variable (the table in Figure 8.4)
T_s = schedule date or duration from which the probability is determined
E = expected project duration or mean
V_T = project duration variance

Let's now determine the probability of meeting a scheduled date in our example. The project duration mean is 14.3 days, the variance is 5.27 days, and these results are considered normally distributed. Therefore, if this project were repeated a large number of times under the same conditions, half of the projects would finish in fewer than 14.3 days and the other half would finish in more than 14.3 days. In other words, there is a 50 percent chance that the project duration will be less than 14.3 days and a 50 percent chance that the project will be more. It may be helpful to determine the probability of meeting a different project duration. For example, the probability of finishing the project in 17 days can be determined from Equation 8.5 above and Figure 8.4. The value of z for the cumulative normal distribution is

$$\Pr \{Z \le z = [17 - 14.3] \div \sqrt{5.27}$$

$$\Pr \{Z \le z = 1.16\} = 0.8770$$

Cumulative Normal
Distribution Table

$\Pr\{z \le 1.65\} = 0.9505$

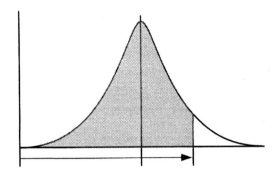

z	0.00	0.01	0.02	0.03	0.04	0.05	0.06	0.07	0.08	0.09
0.0	0.5000	0.5040	0.5080	0.5120	0.5160	0.5199	0.5239	0.5279	0.5319	0.5359
0.1	0.5398	0.5438	0.5478	0.5517	0.5557	0.5596	0.5636	0.5675	0.5714	0.5754
0.2	0.5793	0.5832	0.5871	0.5910	0.5948	0.5987	0.6026	0.6064	0.6103	0.6141
0.3	0.6179	0.6217	0.6255	0.6293	0.6331	0.6368	0.6406	0.6443	0.6480	0.6517
0.4	0.6554	0.6591	0.6628	0.6664	0.6700	0.6736	0.6772	0.6808	0.6844	0.6879
0.5	0.6915	0.6950	0.6985	0.7019	0.7054	0.7088	0.7123	0.7157	0.7190	0.7224
0.6	0.7258	0.7291	0.7324	0.7357	0.7389	0.7422	0.7454	0.7486	0.7518	0.7549
0.7	0.7580	0.7612	0.7642	0.7673	0.7704	0.7734	0.7764	0.7794	0.7823	0.7852
0.8	0.7881	0.7910	0.7939	0.7967	0.7996	0.8023	0.8051	0.8078	0.8106	0.8133
0.9	0.8159	0.8186	0.8212	0.8238	0.8264	0.8289	0.8315	0.8340	0.8365	0.8389
1.0	0.8413	0.8438	0.8461	0.8485	0.8508	0.8531	0.8554	0.8577	0.8599	0.8621
1.1	0.8643	0.8665	0.8686	0.8708	0.8729	0.8749	0.8770	0.8790	0.8810	0.8830
1.2	0.8849	0.8869	0.8888	0.8907	0.8925	0.8944	0.8962	0.8980	0.8997	0.9015
1.3	0.9032	0.9049	0.9066	0.9082	0.9099	0.9115	0.9131	0.9147	0.9162	0.9177
1.4	0.9192	0.9207	0.9222	0.9236	0.9251	0.9265	0.9279	0.9292	0.9306	0.9319
1.5	0.9332	0.9345	0.9357	0.9370	0.9382	0.9394	0.9406	0.9418	0.9429	0.9441
1.6	0.9452	0.9463	0.9474	0.9484	0.9495	0.9505	0.9515	0.9525	0.9535	0.9545
1.7	0.9554	0.9564	0.9573	0.9582	0.9591	0.9599	0.9608	0.9616	0.9625	0.9633
1.8	0.9641	0.9649	0.9656	0.9664	0.9671	0.9678	0.9686	0.9693	0.9699	0.9706
1.9	0.9713	0.9719	0.9726	0.9732	0.9738	0.9744	0.9750	0.9756	0.9761	0.9767
2.0	0.9772	0.9778	0.9783	0.9788	0.9793	0.9798	0.9803	0.9808	0.9812	0.9817
2.1	0.9821	0.9826	0.9830	0.9834	0.9838	0.9842	0.9846	0.9850	0.9854	0.9857
2.2	0.9861	0.9864	0.9868	0.9871	0.9875	0.9878	0.9881	0.9884	0.9887	0.9890
2.3	0.9893	0.9649	0.9898	0.9901	0.9904	0.9906	0.9909	0.9911	0.9913	0.9916
2.4	0.9919	0.9920	0.9922	0.9925	0.9927	0.9929	0.9931	0.9932	0.9934	0.9936
2.5	0.9938	0.9940	0.9941	0.9943	0.9945	0.9946	0.9948	0.9949	0.9951	0.9952
2.6	0.9953	0.9955	0.9956	0.9957	0.9959	0.9960	0.9961	0.9962	0.9963	0.9964
2.7	0.9965	0.9966	0.9967	0.9968	0.9969	0.9970	0.9971	0.9972	0.9973	0.9974
2.8	0.9974	0.9975	0.9976	0.9977	0.9977	0.9978	0.9979	0.9979	0.9980	0.9981
2.9	0.9981	0.9982	0.9982	0.9983	0.9984	0.9984	0.9985	0.9985	0.9986	0.9986
3.0	0.9987	0.9987	0.9987	0.9988	0.9988	0.9989	0.9989	0.9989	0.9990	0.9990

FIGURE 8.4 Cumulative normal distribution

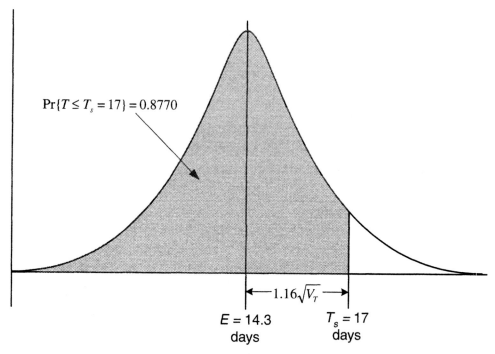

$$\Pr\{T \le T_s = 17\} = 0.8770$$

$$1.16\sqrt{V_T}$$

$E = 14.3$
days

$T_s = 17$
days

FIGURE 8.5   Distribution of probability that $T_s = 17$ or fewer days

Referring to Figure 8.4, a z value 1.16 yields a probability of 0.8770, or about 88 percent. So for the data given, there's an 88 percent chance that the project will finish in 17 or fewer days. The probability of 0.8770 is actually the area under the standard normal curve (with a total area of 1.000), as can be seen in Figure 8.5. The z value of 1.16 indicates that the scheduled time, $T_s = 17$ days, is 1.16 standard deviations greater than the expected duration, $E = 14.3$ days. If we wanted to know the duration that yields a 95 percent chance of finishing the project, we would determine the corresponding z value from Figure 8.4—1.645 (interpolated)—and plug it into Equation 8.6 as follows:

$$T_s = \left[\sqrt{5.27} \times 1.645 \right] + 14.3 = 18.1 \text{ days}$$

Thus, there is a 95 percent chance that the project will finish in 18.1 days or less. This can be taken a step further to determine the probability of a project taking more than a particular number of days. Using the results thus far, we know that if there is an 88 percent chance of completing the project in 17 days or fewer, then there is a 12 percent (100 - 88) chance that it will take more than 17 days. Also, if, there's a 95 percent chance of finishing the project in 18.1 days, then there's a 5 percent (100 - 5) chance that it will take more than 18.1 days to complete the project.

The PERT statistical approach is seldom used in the construction industry today. As noted earlier, the process was devised for use in a research and development project, the Polaris missile system, in the 1950s and still remains mostly applicable today to R&D projects that are very costly and time-consuming and are under tight time/cost constraints. But because the PERT method is one that utilizes an intriguing approach to manage uncertainty in activity durations, it is worth our time to study it in order to gain more understanding of duration uncertainty.

8.5   SUMMARY

The network model is the tangible result of the network plan used for management purposes and, applied through project scheduling, is only a representation of the actual construction process. The performance of the model is only as good as the data used for its generation. In CPM and PN modeling, time estimates for activity durations are fixed and known. Uncertainty in activity duration estimates is undeniable, particularly during the planning stages of a project. The program evaluation and review technique (PERT) is the most common formal method that considers time estimate uncertainty and is mostly applied in projects where considerable time variations are expected for activity durations.

Uncertainty in activity duration estimates due to time/resource constraints may be due to adverse or unexpected weather conditions, low-productivity or maintenance problems with equipment, and/or a high degree of uncertainty in unique construction methods. The PERT procedure provides a probability that a project will be completed on or before a specified completion date based on variable time estimates of activity durations. PERT utilizes an average, or mean, time and a variance for each activity, calculated from a set of user-provided values. The procedure assumes that activity duration estimates are independent random variables, which tend to follow the central limit theorem. This theorem can be used to describe the behavior of the time estimates, their means, and the associated variations and their effect on project scheduling computations. The PERT procedure allows the estimation of the probability of completing a project on time (or within certain time windows) with uncertain time estimates present.

CHAPTER 8   QUESTIONS/PROBLEMS

1. Team Activity –Research the PERT procedure, detailing its developmental history and major uses today.
2. Explain the reasons for the limited application of the PERT procedure in the construction industry.
3. Describe the reasons for and the results of uncertainty in the estimation of project activity time duration.

4. The following data are given for a construction project:

ACTIVITY	IMMEDIATE PREDECESSOR	A	M	B
A	—	2	3	5
B	A	2	4	7
C	A	4	5	7
D	B, C	3	5	7
E	C	2	5	9
F	D, E	3	4	7
G	F	5	7	12

a. Draw an AON diagram of the project.
b. Calculate the t_i and V_i of each activity and develop a forward and backward pass for the project to determine activity ES, EF, LS, LF, and float; expected project duration (E); and total project V_T. Show these values clearly on your AON diagram.
c. Determine the probability of the project finishing in:

- 27 or fewer days
- More than 26 days
- 30 or fewer days

REFERENCES

See References on page 199.

APPENDIX

THREE-UNIT TOWNHOUSE

This appendix contains the following information:

- ❑ Narrative description
- ❑ Layout drawings
- ❑ Project activity list/durations
- ❑ Work breakdown structure
- ❑ Project schedule—network diagrams:
  - AON diagram—FS relationships only
  - PN diagram—FS, SS, FF relationships

NARRATIVE DESCRIPTION

The three-unit townhouse is a two-story structure approximately 28 ft by 60 ft with two 1,125 ft^2 townhouse units (at approximately 18 ft by 28 ft per floor) and one 1,350 ft^2 unit (at approximately 24 ft by 28 ft). The layouts of the townhouses are relatively simple, with a living area, kitchen, and half bath (and extra bedroom for the larger unit) on the first floor and two bedrooms (three for the larger unit) and full bath on the second floor. The structure is made of traditional materials using typical building techniques—wooden stud platform framing, sloped roof, and siding. Unlike a traditional residential house, firewalls are placed between the units, and other mechanical/electrical services are similarly located and shared where possible. Further, major building steps such as foundation and framing will be fabricated for the entire structure (all three townhouses) rather than separately or sequentially for single units.

Townhouse front elevation

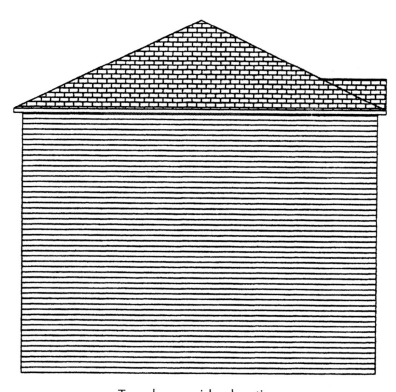

Townhouse side elevation

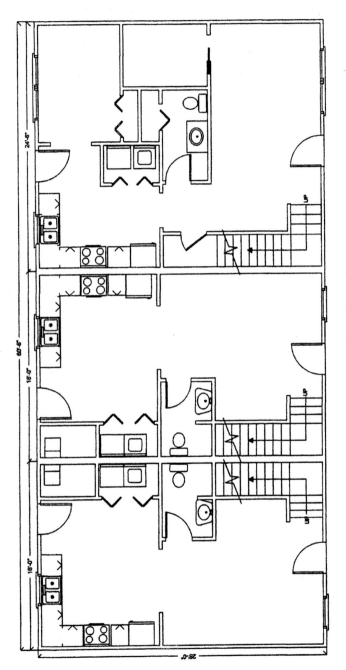

Townhouse first-floor plan

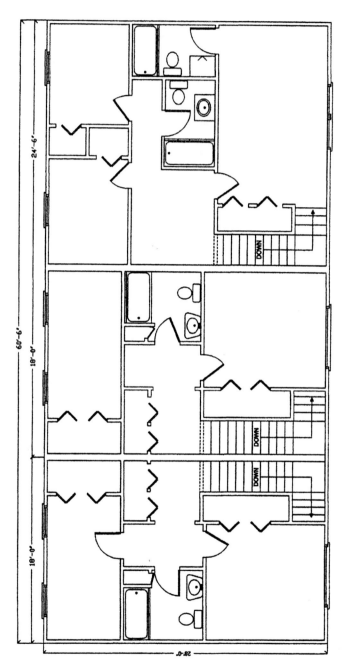

Townhouse second-floor plan

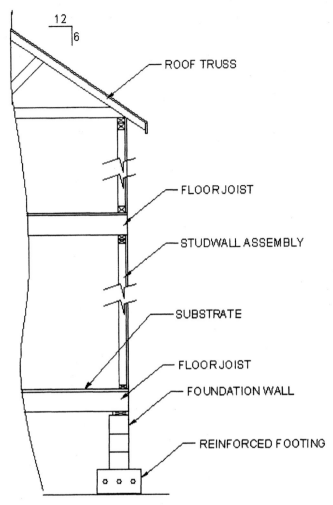

12
6

— ROOF TRUSS

— FLOOR JOIST

— STUDWALL ASSEMBLY

— SUBSTRATE

— FLOOR JOIST

— FOUNDATION WALL

— REINFORCED FOOTING

Townhouse wall section

Three-Unit Townhouse Activity List

1000 Site Work — Duration (days)

1100	Site Layout and Engineering	1
1200	Access Roads	1
1300	Clear the Site	1
1400	Rough-in Building Location	1
1500	Grade and Drain	1
1600	Finish Grade	2

2000 Footings and Foundations

2100	Building Layout and Engineering	1
2200	Excavate Footings	1

2300	Formwork and Reinforcing Steel	2
2400	Place and Cure Concrete	1
2500	Strip Forms	1
2600	Masonry	2
2700	Damp-proofing	1

3000 Erect Structure/Frame

3100	Building Layout and Engineering	1
3200	Build All Floors	7
3300	Build All Walls	10
3400	Build Porches and Decks	2
3500	Build Roof System	12
3600	Place Exterior Doors and Windows	2

4000 Mechanical Systems

4100	HVAC System Rough-in	6
4200	Plumbing Rough-in	7
4300	Electrical Rough-in	7
4400	Miscellaneous Systems Rough-in	2

5000 Interior Finish

5100	Insulation	3
5200	Sheetrock	8
5300	Interior Doors	2
5400	Trim Carpentry	8
5500	Paint and Wall Coverings	4
5600	Cabinetry	2
5700	Floor Coverings	4
5800	Hardware	3
5900	Mechanical Finish	3

6000 Exterior Finish

6100	Roofing System	4
6200	Siding	5
6300	Masonry	3
6400	Painting	2
6500	Landscaping	4

7000 Inspection and Punch List

7100	Inspection by Contractor	2
7200	Punch list	5
7300	Close-out inspection	1

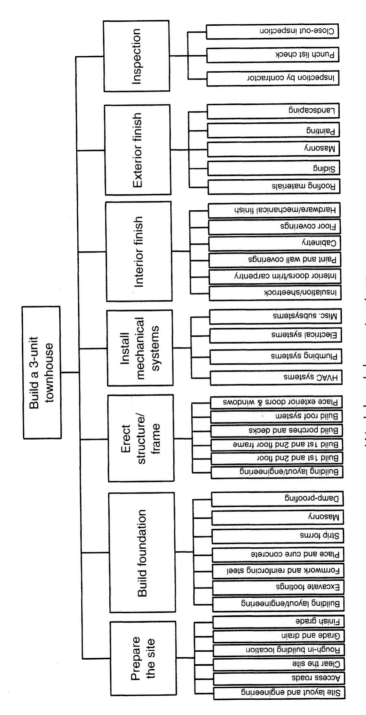

Work breakdown structure

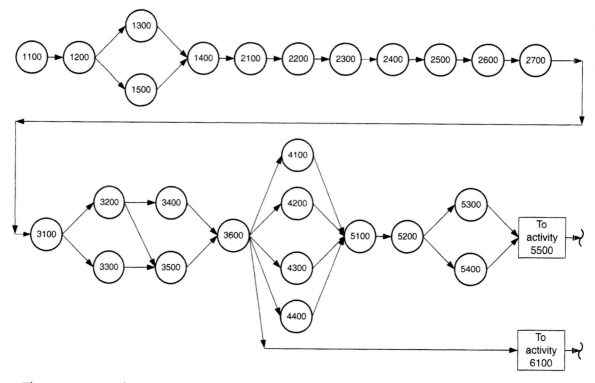

Three-unit townhouse network (AON) diagram—FS schedule (All activity relationships are finish-to-start relationships)

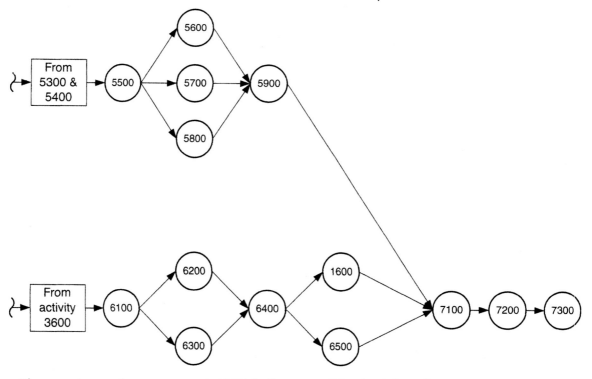

Three-unit townhouse network (AON) diagram—FS schedule (All activity relationships

are finish-to-start relationships)

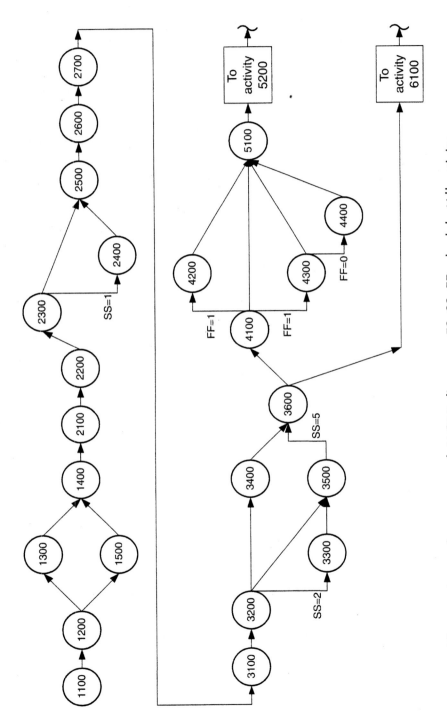

Three-unit townhouse network (AON) diagram—FS, SS, FF schedule (All activity relationships are finish-to-start relationships unless otherwise indicated)

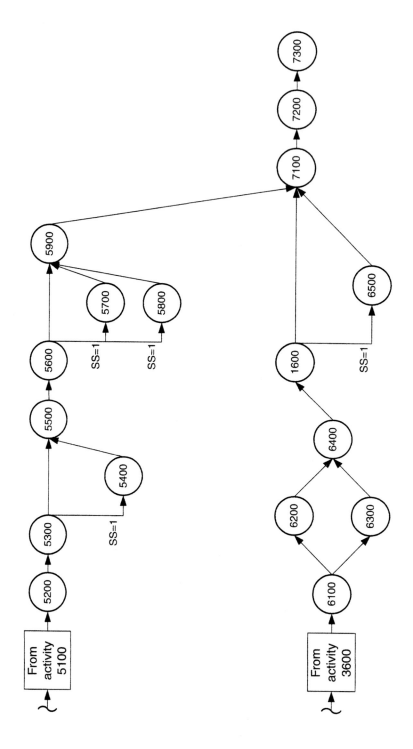

Three-unit townhouse network (AON) diagram—FS, SS, FF schedule (All activity relationships are finish-to-start relationships)

REFERENCES

Adamiecki, Karol. 1931. Harmonygraph. *Przeglad Organizacji (Polish Journal on Organizational Review)*.

Burgess, A. R., and J. B. Killebrew. 1962, March/April, "Variation in Activity Level on a Cyclic Arrow Diagram." *Journal of Industrial Engineering*. Vol. 13, No. 2.

Feigenbaum, Leslie. 1998. *Construction scheduling with primavera project planner*. Upper Saddle River, NJ: Prentice Hall.

Feigenbaum, Leslie. 2001. *Construction scheduling with primavera project planner*, 2d ed. Upper Saddle River, NJ: Prentice Hall.

Fondahl, John W. 1962. *A non-computer approach to the critical path method for the construction industry*. 2d ed. (Stanford, CA: Department of Civil Engineering, Stanford University).

Harris, R. B. 1990. "Packing Method of Resource Leveling (Pack)." *Journal of Construction Engineering and Management*, Vol. 116, No. 2, pp. 331–350.

Hendrickson, C., and T. Au. 1989. *Project management for construction*. Englewood Cliffs, NJ: Prentice Hall.

Hinze, Jimmie W. 1998. *Construction planning and scheduling*. Upper Saddle River, NJ: Prentice Hall.

IBM,1968. *Project Management System, Application Description Manual*, (H20–0210).

Kerzner, Harold. 1999. The growth of modern project management. *Project Management Journal*, 25, no. 2 (June): 6–8.

Lehner, Mark. 1997. *The complete pyramids: Solving the ancient mysteries*. London: Thames & Hudson.

Levy, F. K., G. L. Thompson, and J. D. Wiest. 1962. "Multiship, Multishop, Workload-Smoothing Program." *Naval Research Logistics Quarterly*, Vol. 9, No. 1, pp. 37–44.

Liberatore, M. J., B. Pollack-Johnson, and C. A. Smith. 2001. "Project management in construction: Software use and research directions." *Journal of Construction Engineering and Management*. (March/April): 101–107.

Mahoney, William D. 1990. *Means estimating handbook*. Kingston, MA: R. S. Means Company.

Marchman, David A. 2000. *Scheduling with SureTrak*. Albany, NY: Delmar-Thompson Learning.

Martinez, J., and P. Ioannou. 1993. "Resource Leveling Based on the Modified Minimum Moment Heuristic." *Proceedings of the 5th International Conference on Computing in Civil and Building Engineering*. Reston, VA: ASCE, pp. 287–294.

Moder, J. J., C. R. Phillips, and E. W. Davis. 1983. *Project management with CPM, PERT, and Precedence Diagraming*, 3d ed. New York: Van Nostrand Reinhold.

Naylor, Henry. 1995. *Construction project management: Planning and scheduling*. Albany, NY: Delmar Publishers.

Project Management Institute. 2001. "Practice Standard for Work Breakdown Structures." Newton Square, PA: Project Management Institute.

Savin, D., S. Alkass, and P. Fazio. 1996. "Construction Resource Leveling Using Neural Networks." *Canadian Journal of Civil Engineering*, Vol. 23, No. 3, pp.917–925.

Siddens, Scott (ed.). 1999. *Walker's building estimator's reference book*. Lisle, IL: Frank R. Walker Company.

Spinner, M. Pete. 1997. *Project management: Principles and practices*. Upper Saddle River, NJ: Prentice Hall.

U.S. Bureau of Labor Statistics. 2000–2001. *Occupational Outlook Handbook* (Bulletin 2520). Chicago: Publications Sales Center. http://stats.bls.gov/oco/ocos005.htm.

Wideman, R. M. 1986. "The PMBOK Report—PMI Body of Knowledge Standard," *Project Management Journal* 17, no. 3 (August): 15–24.

Wiest, J. D., and F. K. Levy. 1977. *A Management Guide to PERT/CPM*, 2d ed. Englewood Cliffs, NJ: Prentice Hall.

INDEX